BROADBAND POWER-LINE COMMUNICATION SYSTEMS

BROADBAND POWER-LINE COMMUNICATION SYSTEMS

THEORY & APPLICATIONS

J. Anatory

University of Dodoma, Tanzania

&

N. Theethayi

Bombardier Transportation, Sweden

WIT*PRESS* Southampton, Boston

J. Anatory
University of Dodoma, Tanzania

N. Theethayi
Bambardier Transportation, Sweden

Published by

WIT Press
Ashurst Lodge, Ashurst, Southampton, SO40 7AA, UK
Tel: 44 (0) 238 029 3223; Fax: 44 (0) 238 029 2853
E-Mail: witpress@witpress.com
http://www.witpress.com

For USA, Canada and Mexico

WIT Press
25 Bridge Street, Billerica, MA 01821, USA
Tel: 978 667 5841; Fax: 978 667 7582
E-Mail: infousa@witpress.com
http://www.witpress.com

British Library Cataloguing-in-Publication Data
A Catalogue record for this book is available
from the British Library

ISBN: 978-1-84564-416-1

Library of Congress Catalog Card Number: 2009940061

The texts of the papers in this volume were set individually by the authors or under their supervision.

Printed in Great Britain by Quay Digital, Bristol.

Contents

Chapter 6
Modulation and coding techniques for power-line communications systems

Chapter 7
Performance of PLC systems that use modulation and coding techniques

Preface

The implementation of access technology is a challenge in both developed and developing countries. There are large investments for installations and maintenance for such access technology networks and power line network is the best candidate. The reason geing, the potential broadband end users first consider having electricity in their household and business areas or within a large scale industry that draws power from electrical grids. Researchers have widely investigated the applicability/feasibility of Power line Network for communication and found that they have enough bandwidth for communication at nearly any data rate. A limitation hindering the communications through such media is the regulations by communication authorities, e.g. in Europe, CENELEC standard have regulated the operation frequencies and maximum power to be transmitted in power line communication (PLC) environment. These led researchers to review PLC systems with regards to frequency band of operations and maximum operating power in various countries, because PLC systems radiate like antenna and can cause interference to other communication medium e.g. wireless, normal broadcast radio and so on. Generally, Power line Networks implemented in different countries have similarities and this makes it easier for PLC modeling. PLC in principle is affected by noise, but this has shown to be similar with other communication channels like cellular and radio and this limits the performance to be achieved. PLC network is divided into three categories, indoor PLC, low voltage PLC, and medium voltage PLC. The low and medium voltage PLC is called access network. Generally, PLC technology can be divided into two groups so called narrowband and broadband technologies. The narrowband technology allows the data rates up to 100kbps while the broadband technology allows data rates beyond 2Mbps. The narrowband services include office and home automation, energy information systems, transportation systems, etc. Currently, there is a growing deployment of PLC technologies in various countries and a number of manufacturers offer PLC products with data rates up to 45Mbps or more.

There are a number of standards activities available today such as IEEE P1675 "Standard for Broadband over Power-line Hardware" a group working on hardware installation and safety issues, IEEE P1775 "Power-line Communication Equipment Electromagnetic Compatibility (EMC) Requirements – Testing and Measurement Methods" a working group focusing on PLC equipment, electromagnetic compatibility requirements, testing and measurement methods, and IEEE P1901 "IEEE P1901 Draft Standard for Broadband over Power line Networks: Medium Access Control and Physical Layer Specifications". These standards of course serve as guidelines for PLC system design.

Currently, there are few books covering PLC technologies, the first book by Klaus Dostert covering mainly narrowband PLC technology, focusing on transmission aspects at narrowband frequencies. The second book by Halid Hrasnica, Abdelfatteh Haidine and Ralf Lehnert contributes to the design aspects of broadband PLC access systems and their network components, particularly in the medium access control and general aspects on network design. In this book an attempt is made to study power line networks based on transmission line (TL) theory, to understand propagation aspects for channel model development and comparisons, channel capacity and delay spread analyses, methods of incorporating modulation schemes, etc. The authors of this book have been involved in PLC research and through this book we present the subject of PLC keeping both academic and industry audiences in mind, i.e. discussions are largely based on modeling and design.

The topics in this book include classification of BPLC systems, models for analyses based on TL theory, estimation of channel capacity and performance and finally application of modulation, and coding techniques for boosting the performance of BPLC systems. For convenience of the readers a couple of chapters are dedicated to the fundamental aspects of TL, communication and networking theories which act as warm up for other chapters.

During preparation of this book many people were involved to support us in different ways. We would like to thank all those reviewers of IEEE Transaction on Power Delivery, of the IEEE Power Engineering Society (PES), who reviewed the literature in our papers. Those papers are a part of this book. Authors would like to thank colleagues at the College of Informatics and Virtual education, University of Dodoma, Tanzania and Prof. Rajeev Thottappillil of Royal Institute of Technology, Stockholm, Sweden.

Justinian Anatory & Nelson Theethayi

CHAPTER 1

Power-line communications

1 Introduction

Power-line communications (PLC) technology evolved soon after the widespread establishment of electrical power supply distribution systems. We shall concentrate on three distribution systems, namely, medium voltage (MV), low voltage (LV) and indoor voltage (IV) systems. Typical MV systems involve three phase transmission systems of few kilometers long from primary distribution transformer (DT) (high voltage: 3.3/6.6–11/33 kV transmission voltages) to secondary DT operating at few hundreds of volts (LV). The LV systems involve three phase transmission systems of several hundreds of meters long from secondary DT (LV: 400/230 V) to the consumer connections. The IV systems are applicable to transmission systems within the consumer's premises.

In the beginning efforts were to establish low-speed narrowband communication link over power lines. In the year 1922 the first carrier frequency systems began to operate over high-voltage lines in the frequency range 15–500 kHz for telemetry applications (still working today) as documented by Broadridge [1] and Dostert [2]. In the 1930s, ripple carrier signaling was introduced on the MV and LV distribution systems [1]. Later, the narrowband communication was used for applications such as control and telemetry of electrical equipments like remote meter reading, dynamic tariff control, load management, load profile recording, etc. as outlined by Newbury [3, 4]. There is yet another application of narrowband technology referred to as distribution line carrier or high-speed narrowband PLC. It uses frequency range of 9–500 kHz with data rate up to 576 kbps, typically used for multiple real-time energy management applications, like supervisory control and data acquisition (SCADA), automatic meter reading (AMR) and power quality monitoring system. A narrowband PLC channel model and a survey on such communications can be found in Cooper & Jeans [5].

Broadband in general electronics and telecommunications language is a term which refers to a signaling method involving a relatively wide range of frequencies that may be divided into channels or frequency bins. PLC is also called mains communication, power-line telecommunications (PLT), power band or power-line networking (PLN) which simply describes using power distribution wires for both power delivery and distribution of data. Broadband over power-lines (BPL), also known as power-line Internet or power band is the use of PLC technology to provide broadband Internet access through ordinary power lines. This means high-speed Internet access to the computer or communication devices is available by just plugging a BPL 'modem' to any power outlet in a BPL-equipped building. BPL offers a number of benefits over regular cable or digital subscriber line (DSL) connections. In addition to providing people in different locations with Internet access, it is perhaps easier to provide digital entertainment services using television.

However, variations in the load impedances and topology of the electricity networks means providing services is far from being a standard. Further, repeatable processes and the bandwidth of a BPL system can provide services better compared to cable and wireless systems as discussed in Homeplug 2006 [6]. The PLC can be used for remote monitoring and control of systems, sensor control and data acquisition, voice-over-power-line and multimedia applications such as e-learning, e-health, etc. irrespective of whether it is used in home/building or industrial premises [7–15, 29–35]. Some special applications out of many are:

- *Transportation systems:* PLC technology enables in-vehicle network communication of voice, data, music and video signals by digital means over direct current (DC) battery power-line. Some prototypes are in operational for vehicles, using automotive compatible protocols such as CAN-bus, LIN-bus over power-line (DC-LIN) and DC-bus. This scenario is outlined in Maryanka [7]. Automotive applications include mechatronics-based climate control, door module, obstacle detector, etc. [7]. PLC technology is also applicable for onboard high-speed public transport systems, such as a high-speed train, a transoceanic cruise ship, or even a mid- to long-range airplane [7–9, 29, 34].
- *Automatic meter reading:* automatic meter reading (AMR), is the technology of automatically collecting data from energy meter or water metering devices (water, gas and electric) and transferring the data to a central database for billing purpose and/or analyzing. This saves employee (meter reader) trips, and means that billing can be based on actual consumption rather than on an estimate based on previous consumption, giving customers better control of their use of electric energy, gas usage or water consumption [3, 10–12, 26]. Also this technique can be used to control efficiently electrical power theft practice in many countries [10].
- *Smart grid:* smart grid is an integrated application of information and communication technology (ICT) on electric power transmission and distribution networks. A smart grid replaces analog electromechanical meters with digital meters that record real-time usage data. Smart meters provide a communication path extending from generation plants to electrical outlets (smart sockets) by using robust two-way communications, advanced sensors and distributed computing techniques to improve the efficiency, reliability and safety of power delivery and economy in electricity usage [12–15, 21, 25–28].
- *Radio transmitting programs:* PLC was and is used to transmit radio programs over power-lines or over telephone lines, for example in Germany and Switzerland, the system were called as 'Drahtfunk' and 'Telefonrundspruch' respectively [19, 25]. In all cases the radio programs were fed by special transformers into the lines. In order to prevent uncontrolled propagation, filters for the carrier frequencies of the PLC systems were installed in substations and at line branches [19, 25].

All the above applications depend on the available channel capacity over the power lines in the particular frequencies of usage. In this book we shall investigate through transmission line theory (discussed in the next chapter) as to how well we can use the power lines for broadband power-line communications (BPLC) applications.

1.1 Topology and components used PLC systems

In some countries typical power-line network topology is as shown in Fig. 1; the line length from far end users to DT is about 1.2 km (LV network), and from DT to primary substation is about 4 km (MV network). The maximum number of customers per phase is about 70 and the maximum number of DTs is about 20 to primary substation [11]. The proposed data network configuration from customer premises to DTs is shown in Fig. 2. In Fig. 2, different sub-networks N1, N2 and N3 which consist of different households are connected to the high-voltage side using bridging routers (R1–R3) located at DTs.

Each sub-network connect consists of about 210 end users for three phase systems. The bridging routers bypass the DTs (connected directly to high-voltage side) and service router (SR) is for network reliability. The primary gateway is connecting the PLC network in a given area to other networks. It receives and sends broadband data from or to bridging routers and other networks. Figure 3

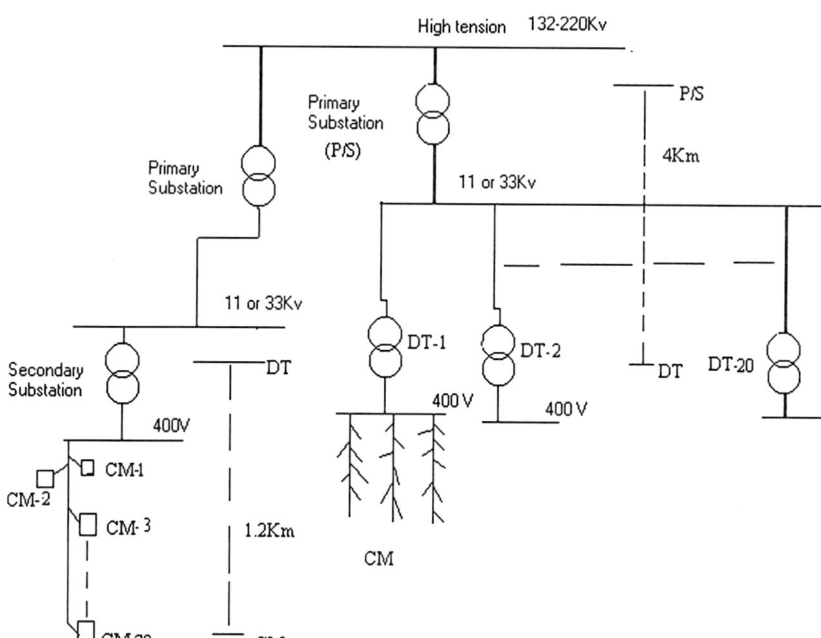

Figure 1: A typical power-line network from customer's location to primary substation adapted from Ref. [10, 11].

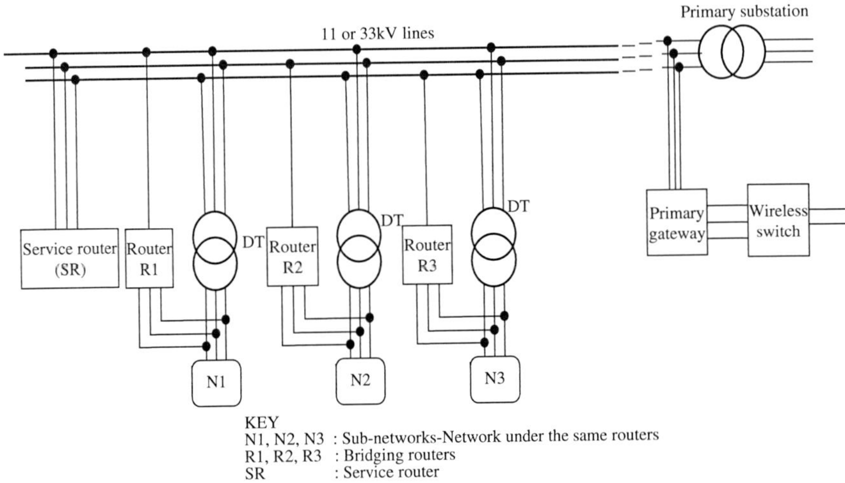

Figure 2: Layout of PLC network interconnections showing different routers located at DTs [11].

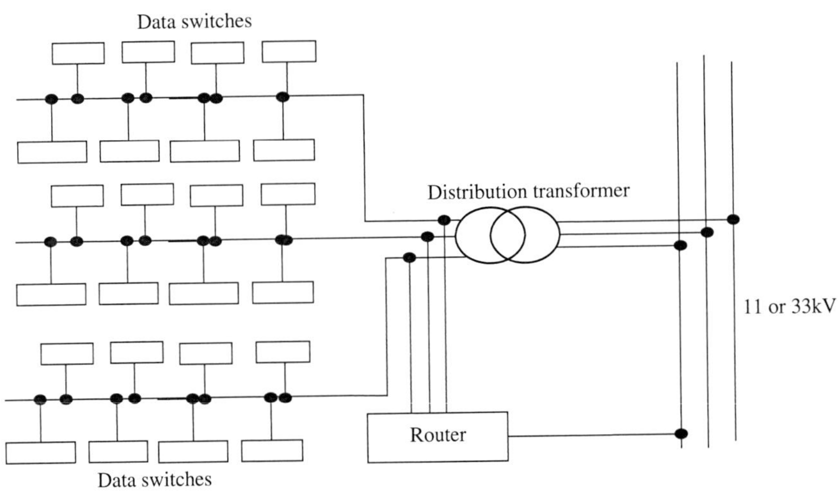

Figure 3: Power-line network layout from customer premises to bridging router at DT [11].

shows data switches (at customer premises) communicating with the primary router through a bridging router at DT. The combinations of data switches at customer premises form the sub-networks (N1–N3). Figure 4 is a simplified indoor BPLC home networking which connects the inside building equipments through data switches to LV access network. The data switches, computers and all

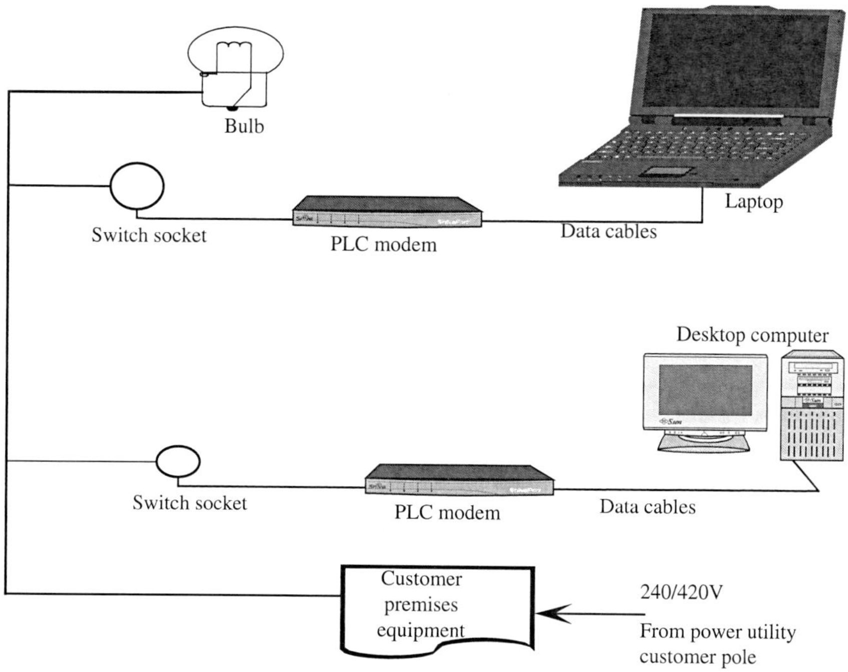

Figure 4: Layout of PLN in a customer premises.

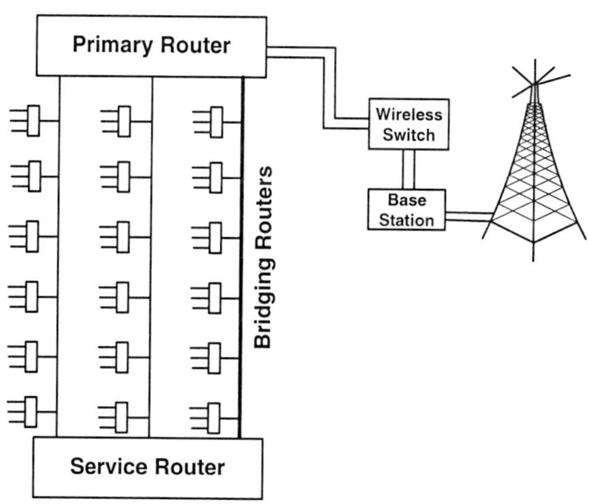

Figure 5: Networking substations with wireless network [11].

communications equipments are connected to electricity networks through coupling circuits (adapter). Figure 5 shows the BPLC network connections with other communication systems such as wireless networks. This connects wireless network at a base station which is located at a primary router [11].

2 Standardization and research group activities

Several competing standards are evolving as indicated below:

- *European Telecommunications Standards Institute (ETSI) power-line telecommunications (PLT):* this project provides a necessary standards and specifications for voice and data services over the power line transmission and distribution network and/or in-building electricity wiring. The standard discusses interoperability aspects between equipment from different manufacturers and co-existence of multiple power-line systems within the same environment [16].
- *Home-Plug Power-Line Alliance*: The Home Plug Power-Line Alliance is a global organization consisting of some 65 member companies. Their mission is to enable and promote rapid availability, adoption and implementation of cost-effective, interoperable and standards-based home power-line networks and products. Because Home Plug technology is based on the contributions of multiple companies from around the world, the resulting standards are expected to offer best performance. The Home Plug Power-Line Alliance has defined some standards like, (a) Home Plug 1.0 – specification for connecting devices via power-lines in the home, (b) Home Plug AV – designed for transmitting high definition television (HDTV) and VoIP around the home, (c) Home Plug BPL – a working group to develop a specification for to-the-home connection and (d) Home Plug Command and Control (CC) – command and control a specification to enable advanced, whole-house control of lighting, appliances, climate control, security and other devices [17–19].
- *Institute of Electrical and Electronics Engineers (IEEE)*: the standards are due to the IEEE BPL Study Group. Some of those standards are: (a) IEEE P1675 'Standard for Broadband over Power-line Hardware' is a working group working on hardware installation and safety issues; (b) IEEE P1775 'Power-Line Communication Equipment – Electromagnetic Compatibility (EMC) Requirements – Testing and Measurement Methods' is a working group focused on PLC equipment, EMC requirements and testing and measurement methods; (c) IEEE P1901 'IEEE P1901 Draft Standard for Broadband over Power-Line Networks: Medium Access Control and Physical Layer Specifications' is a working group for delivering BPL. The aim is to define medium access control and physical layer specifications for all classes of BPL devices – from long distance connections to those within subscriber premises [17–19, 23].
- *POWERNET*: this is a research and development project with funding from the European Commission. It aims at developing and validating a 'plug and play' cognitive broadband over power-lines (CBPL) communications equipment that meet the regulatory requirements concerning electromagnetic radiations and can deliver high data rates while using low transmit power spectral density and working at low signal-to-noise ratio [20, 21].

- *Open PLC European Research Alliance*: Open PLC European Research Alliance (OPERA) is a research and development project with funding from the European Commission. It aims at improving/developing PLC services and system standardization [22, 30].
- *Universal Power-Line Association (UPA)*: The UPA aligns industry leaders in the global PLC market to ensure deployment of interoperable and coexisting PLC products to the benefit of consumers worldwide [23, 24].

References

[1] Broadridge, R., Power-line modems and networks. *IEE 4th International Conference on Metering Applications and Tariffs for Electricity Supply IEE Conf. Publ.*, London, UK, pp. 294–296, 1984.

[2] Dostert, K., Telecommunications over the power distribution grid – possibilities and limitations. *Proceedings of the International Symposium on Power-Line Communications and Its Applications*, Essen, Germany, pp. 1–9, 2nd–4th April 1997.

[3] Newbury, J., Metering Communications and Services Using the Low Voltage Communications, Power Communications Research Group, Open University Manchester, England, 1998 online http://www.metering.com/archieve/973/30_1.html

[4] Newbury, J., Communication requirements and standards for low voltage mains signaling. *IEEE Trans. Power Delivery*, **13(1)**, pp. 46–49, January 1998.

[5] Cooper, D., & Jeans, T., Narrowband, low data rate communications on the low-voltage mains in the CENELEC frequencies – Part I: noise and attenuation. *IEEE Trans. Power Delivery*, **17(3)**, pp. 718–723, July 2002.

[6] Larry, Y., HomePlug AV Technical Overview.*IEEE-ISPLC 2006* – Orlando, FL, USA, 28 March, 2006.

[7] Maryanka, Y., Wiring reduction by battery power line communication. Proc. Passenger Car Elect. Architecture. *IEE Semin.*, pp. 8/1–8/4, June 2000.

[8] Kun-De, L. & Jin-Fu, C., Communications and entertainment onboard a high-speed public transport system. *IEEE Trans. Wireless Commun.* **9(1)**, pp. 84–89, February 2002.

[9] Lienard, M., Carrion, M.O., Degardin, V. & Degauque, P., Modeling and analysis of in-vehicle power line communication channels. *IEEE Transactions on Vehicular Technology*, **57(2)**, pp. 670–679, March 2008.

[10] Anatory, J., Kissaka, M.M., & Mvungi, N.H., PLC network traffic modeling for implementation of remote monitoring of electrical power consumption in Tanzania. *International Symposium on Power Line Communications and Its Applications*, Zaragoza, Spain, 2004.

[11] Anatory, J., Kissaka, M.M. & Mvungi, N.H., Trends in telecommunication services provision: power line network can provide alternative for access in developing countries. *Proc. IEEE Africon*, Gaborone, Botswana, pp. 601–606, September 2004.

[12] Mazza, P., The Smart Energy Network: Electricity's Third Great Revolution. Online http://climatesolutions.org/pubs/pdfs/SmartEnergy.pdf

[13] Voltage Stability Monitoring and Control (VSMC) System in Fujian Power Grid. Proceedings, Power Engineering Society General Meeting, 2007. Online (Tampa, FL, USA: IEEE). 2007-06-24. doi: 10.1109/PES.2007.385975. Lay summary.

[14] Smart Grid Working Group. Challenge and Opportunity: Charting a New Energy Future, Appendix A: Working Group Reports. Energy Future Coalition. Online http://www.energyfuturecoalition.org/files/webfmuploads/EFC_Report/EFC

[15] Smartgrids Advisory Council. Driving Factors in the Move Towards Smartgrids. European Smartgrids Technology Platform: Vision and Strategy. European Commission. p. 9. ISBN 92-79-01414-5.Online www.smartgrids.eu/documents/vision.pdf

[16] ETSI, www.etsi.org/WebSite/Technologies/Powerline.aspx

[17] Galli, S., the Inter Inter-Co-PHY Protocol (IPP): A Simple Co Existence Protocol. *IEEE International Symposium on Power Line Communications*, Dresden, Germany, 31 March 2009.

[18] Galli, S. & Oleg, L., Recent developments in the standardization of power line communications in the IEEE. *IEEE Communications Magazine*, pp. 64–71, July 2008.

[19] Mains Network, http://en.wikipedia.org/wiki/Mains_network

[20] Sathya, R., Broadband over powerlines: a rapid way to join the knowledge society. *Africa/Middle East Next Generation Network Summit*, Johannesburg, South Africa, 17–19 February 2008. Online www.ist-powernet.org

[21] Sathya, R., Use of ubiquitous powerline infrastructure for providing broadband access to bridge the ICT divide, Brussels, 22 March 2006. Online www.ist-powernet.org

[22] OPERA, Open PLC European research alliance for new generation PLC integrated network phase 2. Online www.ist-world.org/ProjectDetails.aspx

[23] IEEE P1901, Draft Standard for Broadband over Power Line Networks: Medium Access Control and Physical Layer Specifications, PUA and UPA contribution to IEEE 1901 call for submission on BPL access requirements Functional and Technical Requirements, pp. 1–41, February 2006. Online www.upaplc.org

[24] Donald, P., General Overview of the UPA and Its Activities, January 2009. Online www.upaplc.org

[25] http://en.wikipedia.org/wiki/Power_line_communication

[26] http://en.wikipedia.org/wiki/Automatic_meter_reading

[27] http://en.wikipedia.org/wiki/Smart_grid

[28] Tsuzuki, S., High-speed power-line communication and its application to a localization system. *IEICE Transactions on Fundamentals of Electronics Communications and Computer Sciences,* E89-A(11), pp. 3006–3012, 2006.

[29] Anatory, J., Theethayi, N., Thottappillil, R., Kissaka, M.M. & Mvungi, N.H., An experimental validation for broadband power line communication (BPLC) model. *IEEE Transactions on IEEE Power Delivery*, 23(3), pp. 1380–1383, July 2008.

[30] Hrasnica, H., Haidine, A. & Lehnert, R., *Broadband Powerline Communications: Network Design*, John Wiley & Sons Inc., June 2004.

[31] Surekha, T.P., Analysis of effect of power line channel characteristic parameters in broadband power line communication (BPLC) systems, 2009 IEEE/PES Power Systems Conference and Exposition, 03/2009.

[32] Tao, Z., Xiaoxian, Y. & Baohui, Z., Broadband transmission characteristics for power-line channels. *IEEE Transactions on Power Delivery*, 21(4), pp. 1905–1911, 2006.

[33] Anatory, J., Theethayi, N., Thottappillil, R., Kissaka, M.M. & Mvungi, N.H., Expressions for current/voltage distribution in broadband power-line communication networks involving branches. *IEEE Transactions on Power Delivery*, 23(1), pp. 188–195, January 2008.

[34] Kun-De, L., Communications and entertainment onboard a high-speed public transport system. *IEEE Wireless Communications*, 2/2002.

[35] Hosoya, S., OFDM signal transmission characteristics through a switchboard for a power line communication, IEEE International Symposium on Power Line Communications and Its Applications, 03/2009.

CHAPTER 2

Transmission line theory

1 Introduction

This chapter focuses on transmission line theory. The authors wish to emphasize that there is a library of literature (some classical texts worth reading are Paul [1] and Tesche *et al.* [2]) on this subject; hence for brevity we shall focus only on telegrapher's or transmission line equations and the methods by which they can be solved considering lossy transmission line (TL) systems. When signals propagate along a lossy transmission line, wave-propagation effects like signal attenuation and dispersion dominate due to (a) TL type (cable or overhead wire), (b) TL above or below a finitely conducting ground and (c) TL wire material.

2 Transverse electromagnetic waves

Paul [1] mentions that wave propagation on transmission lines, waveguides and antennas are very similar to the so-called uniform plane electromagnetic waves shown in Fig. 1. The term uniform plane waves means that at any point in the space the electric and magnetic field intensity vectors lie in a plane, and the planes at any two different points are parallel. Further, the electric and magnetic field vectors are independent of position in each plane. For discussion of plane waves, assume the magnetic and electric field vectors to lay the xy plane. Note that in Fig. 1, the electric and magnetic fields are perpendicular to each other in the plane under consideration, such a field structure is known as transverse electromagnetic (TEM) field structure. In the TL system we are dealing with, along the length of a normal transmission line, both electric and magnetic fields are perpendicular (transverse) to the direction of wave travel [1]. This mode of wave propagation, can exist only when there are at least two conductors or more out of which one is a return conductor and will be a dominant propagation mode for all frequencies, if the cross-sectional dimensions of the transmission line are small compared to the wavelength of the propagating electromagnetic pulse. Hence, the authors acknowledge that the study presented in this book is valid so

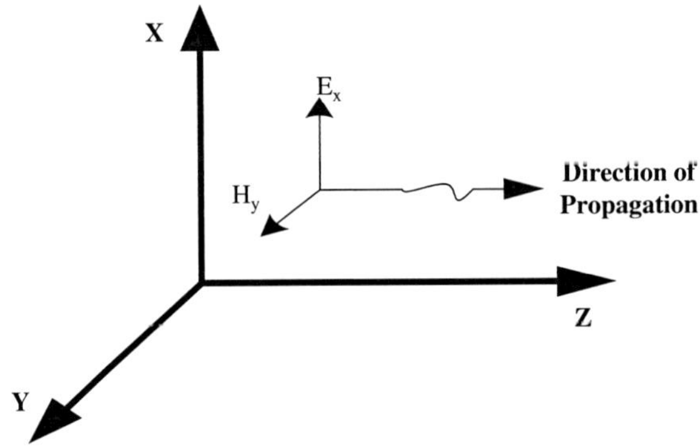

Figure 1: Uniform plane wave: propagation direction and direction of magnetic and electric fields.

long as the TL theory is valid. Let electric field directed along x-axis which is given by (1) be

$$\vec{E} = E_x(z,t) \cdot \vec{a}_x \tag{1a}$$

$$\frac{\partial E_x}{\partial x} = \frac{\partial E_x}{\partial y} = 0 \tag{1b}$$

The TEM field structure necessitates that magnetic field is given by (2).

$$\vec{H} = H_y(z,t) \cdot \vec{a}_y \tag{2a}$$

$$\frac{\partial H_y}{\partial x} = \frac{\partial H_y}{\partial y} = 0 \tag{2b}$$

Using Faraday's and Ampere's law [1], we have the following relationship.

$$\frac{\partial E_x(z,t)}{\partial z} = -\mu \cdot \frac{\partial H_y(z,t)}{\partial t} \tag{3a}$$

$$\frac{\partial H_y(z,t)}{\partial z} = -\sigma \cdot E_x(z,t) - \varepsilon \cdot \frac{\partial E_x(z,t)}{\partial t} \tag{3b}$$

In the above equations, the terms σ, μ and ε are the conductivity, permeability and permittivity of the medium in which the waves are propagating. For the sake of better understanding of the propagation aspects, let us take the discussions entirely in frequency domain. It is well known that the transformation back to time domain can be done by inverse Fourier or Laplace transforms. Transforming the above equations to frequency domain,

$$\frac{dE_x(z,j\omega)}{dz} = -j\omega\cdot\mu\cdot H_y(z,j\omega) \tag{4a}$$

$$\frac{dH_y(z,j\omega)}{dz} = -(\sigma + j\omega\cdot\varepsilon)\cdot E_x(z,j\omega) \tag{4b}$$

If (4a) is differentiated with respect to z and then put into (4b) the wave equations in one dimension are obtained, which are uncoupled second-order differential equations.

$$\frac{d^2E_x(z,j\omega)}{dz^2} = \gamma^2\cdot E_x(z,j\omega) \tag{5a}$$

$$\frac{d^2H_y(z,j\omega)}{dz^2} = \gamma^2\cdot H_y(z,j\omega) \tag{5b}$$

The above equations have the solution as shown in (6), in terms of forward (+) and backward (–) waves.

$$E_x(z,j\omega) = E_m^+\cdot e^{-\gamma\cdot z} + E_m^-\cdot e^{\gamma\cdot z} \tag{6a}$$

$$H_y(z,j\omega) = \frac{E_m^+}{\eta}\cdot e^{-\gamma\cdot z} + \frac{E_m^-}{\eta}\cdot e^{\gamma\cdot z} \tag{6b}$$

In (6), the parameters in front of the exponents are the complex undetermined coefficients and have to be determined from the boundary conditions. The two important coefficients are propagation constant (7a) and intrinsic impedance (7b); from them it can be concluded that when a plane wave propagates in a uniform medium the factors effecting the propagation are the medium and, more importantly, the frequency of the wave itself.

$$\gamma = \sqrt{j\omega\cdot\mu\cdot(\sigma + j\omega\cdot\varepsilon)} = \alpha + j\beta \tag{7a}$$

$$\eta = \sqrt{\frac{j\omega\cdot\mu}{\sigma + j\omega\cdot\varepsilon}} \tag{7b}$$

In any media the wave's propagation characters like the attenuation constant, velocity and the wave length of the wave are determined as in (8a), (8b) and (8c) respectively [1].

$$\alpha = \omega\sqrt{\frac{\varepsilon\cdot\mu}{2}\cdot\left(\sqrt{1+\frac{\sigma^2}{(\omega\cdot\varepsilon)^2}} - 1\right)} \tag{8a}$$

$$v = \frac{\omega}{\beta} = \frac{1}{\sqrt{\frac{\varepsilon \cdot \mu}{2} \cdot \left(1 + \sqrt{1 + \frac{\sigma^2}{(\omega \cdot \varepsilon)^2}}\right)}} \quad (8b)$$

$$\lambda = \frac{2\pi}{\beta} \quad (8c)$$

If the wave is propagating in a lossless media, the conductivity term in the above equations is zero due to which attenuation becomes zero and wave velocity would be the speed of light.

3 Transmission line equations

Consider a two-wire transmission line with the following conditions based on Tesche et al. [2]:

(a) The conductors are parallel to each other and also to the x-axis as shown in Fig. 2
(b) The conductors are perfect and have uniform cross section
(c) The conductors are carrying currents in opposite directions.

Consider the Faraday's law in frequency domain integral form,

$$\int_C \vec{E} \cdot dl = -jw \cdot \mu \cdot \iint_S \vec{H} \cdot ds \quad (9)$$

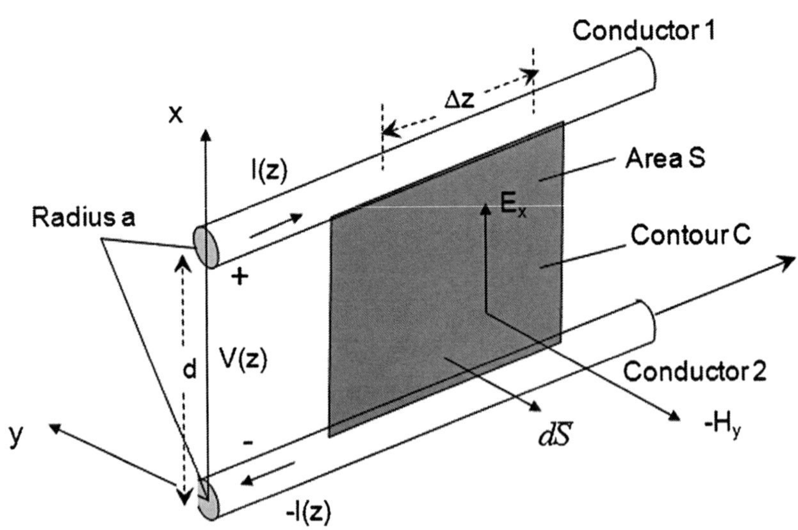

Figure 2: Two-conductor TL system for derivation of first TL equations, adapted from Tesche *et al.* [2].

We integrate in the contour marked C in the anti-clockwise direction and the area marked S, we have,

$$\left\{\begin{array}{l}\int_0^d \left[E_x(x,z+\Delta z)-E_x(x,z)\right]\cdot dx \\ -\int_z^{z+\Delta z}\left[E_z(d,z)-E_z(0,z)\right]\cdot dz\end{array}\right\} = -j\omega\cdot\mu\cdot\int_0^d\int_z^{z+\Delta z} -H_y\cdot dx\cdot dz \qquad (10)$$

In the quasi-static sense, the line-to-line voltage can be defined as (11) with a sign convention such that the voltage on the conductor at level d is positive with respect to the other reference conductor,

$$V(z,j\omega) = -\int_0^d E_x(x,z)\cdot dx \qquad (11)$$

Note that tangential components of electric fields along the conductor are zero for perfect conductors. Dividing (11) by Δz on both sides and taking the limits $\Delta z \to 0$,

$$-j\omega\cdot\mu\cdot\int_0^d H_y(x,z)\cdot dx = -j\omega\cdot L_e\cdot I(z,j\omega) \qquad (12)$$

In the above equation L_e is the external inductance due to flux linkage between the two current carrying conductors. The external impedance is given by $Z_e = j\omega\cdot L_e$.

When the conductors are not perfect, i.e. if they have finite conductivity, then there will be a voltage drop along the conductor due to the component of electric field along the conductor and its frequency penetration into the conductor (skin effect) [3]. This is nothing but Ohm's law [3]. The internal impedance Z_i for circular conductors is given by (13) in terms of Bessel's functions with $\gamma_i = \sqrt{j\omega\cdot\mu_i\cdot\sigma_i}$ [3–5].

$$Z_i = \frac{\gamma_i}{2\pi\cdot\sigma_i\cdot a}\cdot\frac{I_0(\gamma_i\cdot a)}{I_1(\gamma_i\cdot a)} \qquad (13)$$

Note that in (14) the permeability μ_i and conductivity σ_i of the conductor is used. Wedepohl and Wilcox [5] gave an approximate formula for (13) given by (14).

$$Z_i = \frac{\gamma_i}{2\pi\cdot\sigma_i\cdot a}\cdot\coth(0.777a\cdot\gamma_i) + \frac{0.356}{\pi\cdot\sigma_i\cdot a^2} \qquad (14)$$

There is another approximation for (14) proposed by Nahman and Holt [4] and is given by (15).

$$Z_i = \frac{1}{\pi\cdot\sigma_i\cdot a^2} + \frac{1}{2\pi\cdot a}\sqrt{\frac{\mu_i}{\sigma_i}}\cdot\sqrt{j\omega} \qquad (15)$$

Observe that in (15), the first term on right hand side is the resistance of the wire, which means at very low frequencies the resistance of the wire dominates and at high frequencies the total internal impedance increases leading to internal loss. The total impedance of the line is always sum of internal and external impedance $Z = Z_i + Z_e$. Hence the first transmission line equation is given by

$$\frac{dV(z, j\omega)}{dz} + Z \cdot I(z, j\omega) = 0 \tag{16}$$

Now for the derivation of second transmission line equations, consider Fig. 3 – again a two-conductor transmission line scheme. Consider Ampere's law in frequency domain differential form,

$$\nabla \times \vec{H} = j\omega \cdot \vec{\varepsilon} \cdot \vec{E} + \sigma \cdot \vec{E} \tag{17}$$

Applying Stokes theorem, to (17), for a closed surface S_c surrounding one of the conductors as shown in Fig. 3,

$$\oint \vec{H} \cdot dl = j\omega \cdot \varepsilon \cdot \iint_S \vec{E} \cdot ds + \sigma \cdot \iint_S \vec{E} \cdot ds \tag{18}$$

If a contour integration is carried out along the surface of the cylinder, then the result of closed loop integration for the magnetic field in the above equation is difference in the longitudinal current entering and leaving. Further, surface integration of the two terms in the right hand side of the above integration results in

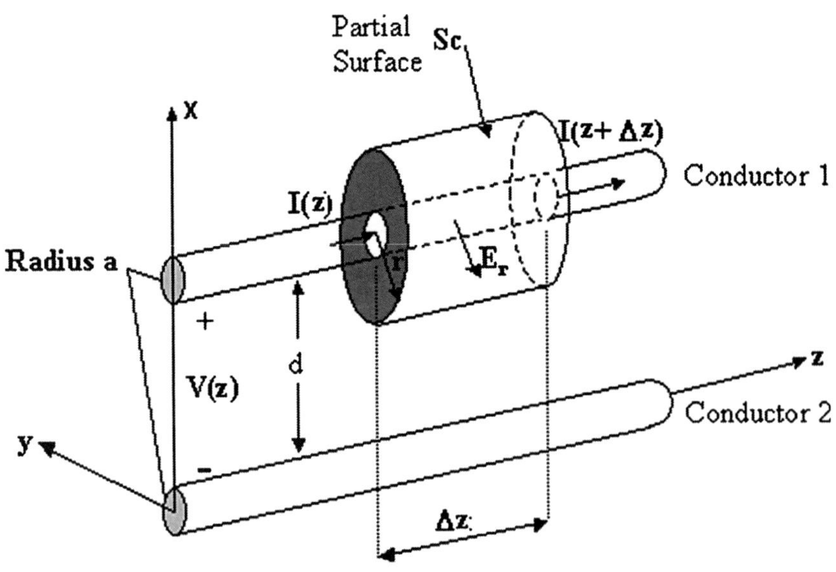

Figure 3: Two-conductor TL system for derivation of second TL equations, adapted from Tesche *et al.* [2].

displacement current due to the permittivity ε of medium and leakage current due to conductivity σ of the medium between the wires shown in Fig. 3.

$$-I(z+\Delta z)+I(z)=j\omega\cdot\varepsilon\iint_{S_c}E_r\cdot r\cdot d\phi\cdot dz+\sigma\iint_{S_c}E_r\cdot r\cdot d\phi\cdot dz \tag{19}$$

The radial electric field in the vicinity of the wire is considered, which will be used for integration in the partial surface S_c. Dividing (21) by Δz and then with limits $r\to a$ and $\Delta z\to 0$ we have

$$\frac{dI(z)}{dz}+j\omega\cdot\varepsilon\cdot\int_0^{2\pi}E_r(z)\cdot a\cdot d\phi+\sigma\int_0^{2\pi}E_r(z)\cdot a\cdot d\phi=0 \tag{20}$$

Assuming that the wire radius is much less than wire separation, the second term and third term in (19) represent the charge and the leakage current terms, which can be represented by the capacitance and conductance terms, respectively, i.e.

$$\varepsilon\cdot\int_0^{2\pi}E_r(z)\cdot a\cdot d\phi=q(z,j\omega)=C_e\cdot V(z,j\omega) \tag{21}$$

$$\sigma\cdot\int_0^{2\pi}E_r(z)\cdot a\cdot d\phi=i_l(z,j\omega)=G_e\cdot V(z,j\omega) \tag{22}$$

The second transmission line equation thus is given by

$$\frac{dI(z,j\omega)}{dz}+G_e\cdot V(z,j\omega)+j\omega\cdot C_e\cdot V(z,j\omega)=0 \tag{23}$$

Similar to impedance in the first transmission line equation, we can define the admittance of the wire as in terms of external conductance as $Y=G_e+j\omega\cdot C_e$. For wires located in free space, the conductance term can be neglected. The second transmission line equation thus is

$$\frac{dI(z,j\omega)}{dz}+Y\cdot V(z,j\omega)=0 \tag{24}$$

It is evident that the two coupled transmission line equations (16) and (24) can be compared to the one-dimensional wave equations as discussed in the previous section. Therefore, we can define for TL systems the propagation constant and the intrinsic or characteristic impedance as

$$\gamma=\sqrt{Z\cdot Y}=\alpha+j\beta \tag{25}$$

$$\eta=\sqrt{\frac{Z}{Y}} \tag{26}$$

The distributed circuit representation of the TL system is shown in Fig. 4.

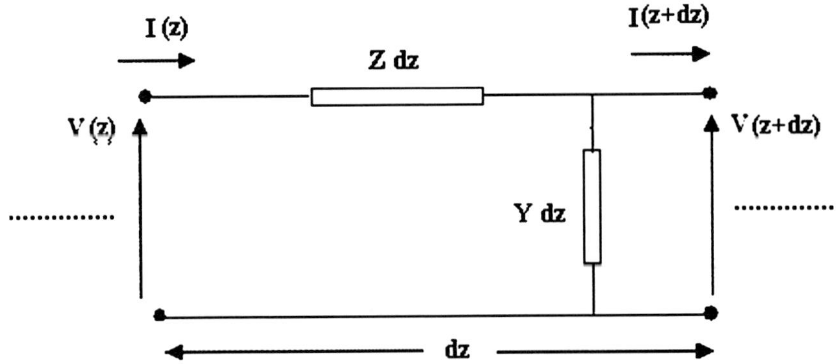

Figure 4: Distributed circuit representation of two-conductor TL systems [1,2].

From the above analysis we have found that it is enough if all the impedance admittance parameters for TL system are found a priory. In this section we shall discuss the useful formula for the transmission line parameters, which is needed for solving TL equations. These parameters are derived from wire geometry, mediums involved and frequency-dependent penetration of electromagnetic fields into the mediums involved [1]. Please refer to standard texts by Paul [1] and Tesche *et al.* [2] for the derivation of those TL parameters. For the case of two wires located in free space like in Fig. 2 or 3, the expressions for inductance in H/m and capacitance in F/m are given by

$$L_e = \frac{\mu}{\pi}\cosh^{-1}\left(\frac{d}{2a}\right) \tag{27}$$

$$C_e = \frac{\varepsilon\pi}{\cosh^{-1}\left(\frac{d}{2a}\right)} \tag{28}$$

Now consider a case when the wire is located at height h above a perfectly conducting ground plane and the currents have their return path via the ground plane; then we can define external inductance and capacitance as

$$L_e = \frac{\mu_0}{2\pi}\ln\frac{2h}{a} \tag{29}$$

$$C_e = \frac{2\pi\varepsilon_0}{\ln\frac{2h}{a}} \tag{30}$$

When the wires are above a finitely conducting ground and the currents are returning through ground, the external impedance due to the electromagnetic fields from

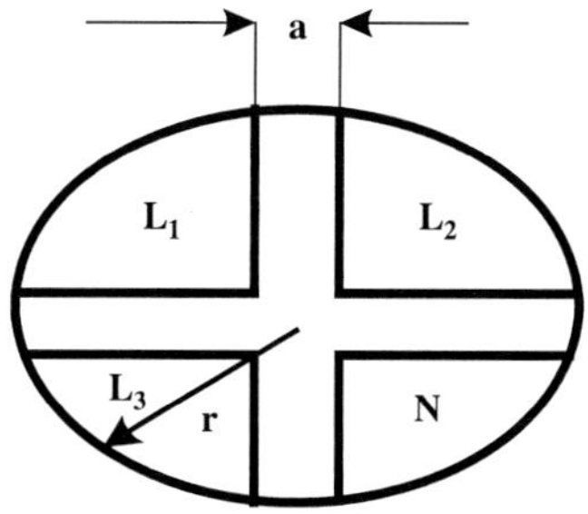

Figure 5: Cross section of a typical power-line cable (four conductors) [7].

wire penetrating into the finitely conducting ground (defined by ground conductivity σ_g and permittivity ε_g) involves a ground impedance term Z_g.

$$Z_e = j\omega \cdot L_e + Z_g \tag{31}$$

The ground impedance expressions for wire located at height h above a finitely conducting ground is given by (32) [6].

$$Z_g = j\omega \cdot \frac{\mu_0}{2\pi} \ln\left[\frac{1+\gamma_g h}{\gamma_g h}\right] \tag{32}$$

The propagation constant of waves in the soil is given by

$$\gamma_g = \sqrt{j\omega\mu_0\sigma_g - \omega^2 \cdot \mu_0 \cdot \varepsilon_g} \tag{33}$$

A typical underground power line cable is shown in Fig. 5; the inductance and capacitance of any line with respect to the neutral line is given by (34) and (35), respectively.

$$L_e = \mu_0 \cdot \mu_r \cdot \frac{a}{r} \tag{34}$$

$$C_e = \varepsilon_0 \cdot \varepsilon_r \cdot \frac{r}{a} \tag{35}$$

Sometimes the multiconductor TL system is used for which the mutual inductance and mutual capacitances have to be found. Whether it is two-conductor system or multiconductor TL system, the analysis for solution of currents and voltages is discussed next. The only difference is that multiconductor TL system involves matrix manipulations, as the voltages and currents on each line are vectors and the impedances and admittances are matrices. The calculations of mutual values are found in Paul [1] and Tesche *et al.* [2].

4 Solution of transmission line equations based on modal analysis based on [1] and [2]

Let us begin with uncoupled second-order TL equations. Note that the impedance matrix and the admittance matrix have elements as applicable based on the discussion in the previous section.

$$\frac{d^2V(x)}{dx^2} = Z \cdot Y \cdot V(x) \tag{36}$$

$$\frac{d^2I(x)}{dx^2} = Z \cdot Y \cdot I(x) \tag{37}$$

Let the total number of conductors excluding the return conductor be n. If we define two $n \times n$ matrices T_{V} and T_{I} which can diagonalize simultaneously both per unit length impedance and admittance matrices, then the solution can be reduced to the solution of n uncoupled first-order differential equations. Thus when the matrices are diagonalized, the system of equations are known as modal equations, which can be easily solved since they are in uncoupled form as in (38) and (39).

$$\frac{dV_{\mathrm{m}}(x)}{dx} = -z \cdot I_{\mathrm{m}}(x) \tag{38}$$

$$\frac{dI_{\mathrm{m}}(x)}{dx} = -y \cdot V_{\mathrm{m}}(x) \tag{39}$$

In the above equations z and y are the modal impedance or admittance matrices (diagonal) and they are connected to the actual line impedance and admittance matrix through the transformation matrix, obtained as in (40) and (41). The second-order modal TL equations in uncoupled form are given by (42) and (43).

$$z = T_{\mathrm{V}}^{-1} \cdot Z \cdot T_{\mathrm{I}} \tag{40}$$

$$y = T_{\mathrm{I}}^{-1} \cdot Y \cdot T_{\mathrm{V}} \tag{41}$$

$$\frac{d^2V_{\mathrm{m}}(x)}{dx^2} = z \cdot y \cdot V_{\mathrm{m}}(x) \tag{42}$$

$$\frac{d^2I_{\mathrm{m}}(x)}{dx^2} = z \cdot y \cdot I_{\mathrm{m}}(x) \tag{43}$$

It is to be noted that $T_I^t = T_V^{-1}$. Consider the second-order modal TL equation corresponding to the current.

$$\frac{d^2 I_m(x)}{dx^2} = T_I^{-1} \cdot Y \cdot Z \cdot T_I \cdot I_m(x) = \gamma^2 \cdot I_m(x) \tag{44}$$

In (44) γ^2 is a diagonal matrix. The solution to the modal currents is given by (45). The exponential terms in (45) are diagonal matrices and other terms are vectors. The final solution for the current is (46).

$$I_m(x) = e^{-\gamma x} \cdot I_m^+ - e^{\gamma x} \cdot I_m^- \tag{45}$$

$$I(x) = T_I \cdot I_m(x) = T_I \cdot \left(e^{-\gamma x} \cdot I_m^+ - e^{\gamma x} \cdot I_m^-\right) \tag{46}$$

The solution to the modal voltages are given by

$$V_m(x) = e^{-\gamma x} \cdot V_m^+ - e^{\gamma x} \cdot V_m^- \tag{47}$$

The final solution for the voltages is (48).

$$V(x) = \left(T_I^{-1}\right)^t \cdot V_m(x) = \left(T_I^{-1}\right)^t \cdot \left(e^{-\gamma x} \cdot V_m^+ + e^{\gamma x} \cdot V_m^-\right) \tag{48}$$

$$V(x) = Z_0 \cdot T_I \cdot \left(e^{-\gamma x} \cdot I_m^+ + e^{\gamma x} \cdot I_m^-\right) \tag{49}$$

$$Z_0 = Z \cdot T_I \cdot \gamma^{-1} \cdot T_I^{-1} = Y^{-1} \cdot T_I \cdot \gamma \cdot T_I^{-1} \tag{50}$$

Now the solutions of voltages and currents at any point on the line can be obtained for a known current or voltage source excitation. The unknown parameters can be obtained by solving the boundary or terminal conditions along the line either using Thevenin or Norton equivalents [1]. For example with resistive loads (Z_S at the source and Z_L at the load end) and with voltage sources at either ends of the line, the following equations are applicable considering the near end and far end boundary conditions.

$$V(0) = V_S - Z_S \cdot I(0) \tag{51}$$

$$V(\ell) = V_L - Z_L \cdot I(\ell) \tag{52}$$

$$\begin{pmatrix} (Z_0 + Z_S) \cdot T_I & (Z_0 - Z_S) \cdot T_I \\ (Z_0 - Z_L) \cdot T_I \cdot e^{-\gamma \ell} & (Z_0 + Z_L) \cdot T_I \cdot e^{\gamma \ell} \end{pmatrix} \cdot \begin{pmatrix} I_m^+ \\ I_m^- \end{pmatrix} = \begin{pmatrix} V_S \\ V_L \end{pmatrix} \tag{53}$$

The authors wish to conclude by saying it is the above TL analyses that will be used in the rest of the chapters, wherever applicable. This chapter was introduced only to convey some TL theory preliminaries.

References

[1] Paul, C.R., *Analysis of Multiconductor Transmission Lines*. John Wiley & Sons Inc., 1994.

[2] Tesche, M.F., Ianoz, V.M. & Karlsson, T., *EMC Analysis Methods and Computational Models*, John Wiley & Sons Inc., 1997.

[3] Schelkunoff, S.A., *Electromagnetic Waves*, D. Van Nostrand Company Inc.: New York, 1943.

[4] Nahman, N.S. & Holt, D.R., Transient analysis of coaxial cables using the skin effect approximation. *IEEE Transactions on Circuit Theory*, **19(5)**, pp. 443–451, 1972.

[5] Wedepohl, L.M. & Wilcox, D.J., Transient analysis of underground power-transmission systems: system-model and wave propagation characteristics. *Proceedings of IEE*, **20(2)**, pp. 253–260, 1973.

[6] Sunde, E.D., *Earth Conduction Effects in the Transmission Systems*, Van Nostrand: New York, 1949.

[7] Anatory, J., Theethayi, N., Thottappillil, R., Kissaka, M. & Mvungi, N., The influence of load impedance, line length, and branches on underground cable power-line communications (PLC) systems. *IEEE Transactions on Power Delivery*, **23(1)**, pp. 180–187, 2008.

CHAPTER 3

Power-line channel models

1 Introduction

Channel models are essential parameters for modeling any communication system. It has been said that optimization of a transmission system is realizable only when a reasonably accurate channel model is available [1] for investigating the power-line network performance in detail. This chapter looks at different power-line channel (PLC) models available in the literature, and compares their accuracy by exact solutions using (the modal analysis as discussed in the previous chapter) time domain simulation software ATP-EMTP [2] wherever applicable. The authors would like to acknowledge that only important channel models are discussed here. Note in all the simulations to follow, a frequency domain solution is first made and then an inverse Fourier transform is made on the frequency response for one-to-one and systematic comparisons of signal propagation characteristics in time domain.

2 Philipps model

The model was proposed by Philipps [3], whose transfer function is given by (1). In (1), *N* is the number of possible signal flow paths, each path delayed by time τ_i is multiplied by a complex factor ρ_i. The parameter ρ_i is the product of transmission and reflection factors and is as indicated in (2). The method developed as described in Ref. [4] did not consider the attenuation during signal propagations. Figure 1 shows the implementation of Philipps echo model with *N* paths.

$$H(f) = \sum_{i=1}^{N} \rho_i e^{-j2\pi f \tau_i} \tag{1}$$

$$\rho_i = |\rho_i| \cdot e^{j\phi_i}, \quad \phi_i = \arctan\left(\frac{\mathrm{Im}(\rho_v)}{\mathrm{Re}(\rho_v)}\right) \tag{2a}$$

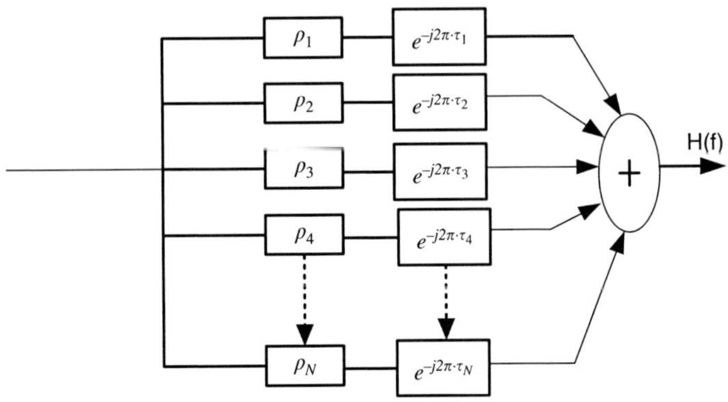

Figure 1: Conceptual sketch of Philipps echo model.

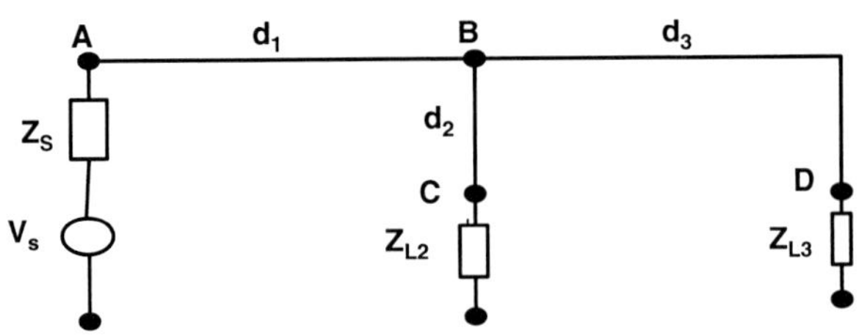

Figure 2: Power-line network with one branch.

$$t_i = \frac{d_i}{v_p} \tag{2b}$$

$$v_p = \frac{c_o}{\sqrt{\varepsilon_r}} \tag{2c}$$

To understand Philipps model for the determination of transfer function as indicated in (1), consider a power-line configuration as given in Fig. 2; the parameters V_s, Z_s, Z_{L2} and Z_{L3} are source voltage, source impedance, load impedance at node C and load impedance at node D, respectively. Assume the signal is injected as shown in Fig. 2, with line lengths as d_1, d_2 and d_3, respectively. Using Philipps echo model the signal propagation is as tabulated in Table 1. The parameters Γ_{13}, ρ_{13}, Γ_{12}, Γ_{23}, ρ_{2c} and ρ_{2B} are the transmission factor from transmission line (TL) 1 to TL 3, the reflection factor from TL 1 to TL 3, the transmission factor from TL 1 to TL 2, the reflection factor at point C and the reflection factor at point B along line 2, respectively. The transmission factors and reflection factors were calculated as in (3)–(6), respectively; in (4) Z_1, Z_2 and Z_3 are characteristic impedances of line 1, line 2 and line 3, respectively. The line characteristic impedances are as in (7).

Table 1: Implementation of Philipps echo model.

Path	Path direction	ρ_i	τ_i
1.	ABD	Γ_{13}	$\frac{d_1+d_3}{v_p}$
2.	ABCBD	$\Gamma_{12}\rho_{2c}\Gamma_{23}$	$\frac{d_1+2\cdot d_2+d_3}{v_p}$
3.	ABCBCBD	$\Gamma_{12}\rho_{2c}^2\rho_{2B}\Gamma_{23}$	$\frac{d_1+4\cdot d_2+d_3}{v_p}$
4.	ABCBCBCBD	$\Gamma_{12}\rho_{2c}^3\rho_{2B}^2\Gamma_{23}$	$\frac{d_1+6\cdot d_2+d_3}{v_p}$
5.	ABCBCBCBCBD	$\Gamma_{12}\rho_{2c}^4\rho_{2B}^3\Gamma_{23}$	$\frac{d_1+8\cdot d_2+d_3}{v_p}$

$$\Gamma_{13}=1+\rho_{13},\ \Gamma_{12}=1+\rho_{12},\ \Gamma_{23}=1+\rho_{23} \tag{3}$$

$$\rho_{13}=\frac{Z_2 // Z_3-Z_1}{Z_2 // Z_3+Z_1},\ \rho_{12}=\frac{Z_2 // Z_3-Z_1}{Z_2 // Z_3+Z_1},\ \rho_{23}=\frac{Z_3 // Z_1-Z_2}{Z_3 // Z_1+Z_2},\ \rho_{2c}=\frac{Z_{L2}-Z_2}{Z_{L2}+Z_2} \tag{4}$$

$$\rho_{2B}=\rho_{23}=\rho_{21} \tag{5}$$

$$\rho_{1B}=\rho_{13}=\rho_{12} \tag{6}$$

$$Z_i=\sqrt{\frac{R+j\omega L_i}{G+j\omega C_i}} \tag{7}$$

$$H(f)=\frac{V_L(f)}{V_A(f)} \tag{8}$$

$$V_L(f)=H(f)\cdot\frac{Z_1}{Z_1+Z_S}\cdot V_S \tag{9}$$

3 Zimmermann and Dostert model

Zimmermann and Dostert [5] developed a channel model to account for the attenuation of the signal flow as given in (10). The concept is similar to Bewley's lattice diagram techniques [6]. Consider Fig. 3, wherein a signal is injected at node A, there will be a propagation of signals towards point B and on reaching at node B, some signals will be reflected and others will be transmitted towards the load at node C. Note that the process is continuous. In (10) each path is characterized by weighting factor g_i which is the product of transmission and reflection factors with path length d_i. The attenuation factor is modeled by the parameters a_0, a_1 and k, which are obtained as data from measurements; also it employs top–down approach. The model was extended to bottom–top model by deriving parameters from actual networks taking into consideration the connected loads [5]. The expressions are as shown in (11)–(13), where $t_{d_i} = \frac{d_i}{v_p}$ and $v_p = \frac{c_o}{\sqrt{\varepsilon_r}}$; also $v_p = \frac{1}{\sqrt{L_e \cdot C_e}}$.

$$H(f) = \sum_{i=1}^{N} g_i e^{-(a_0 + a_1 f^k).d_i} e^{-j2\pi f \frac{d_i}{v_p}} \tag{10}$$

$$H(f) = \sum_{i=1}^{N} g_i''(f) A''(f_i, d) \cdot \exp(-j2\pi f \tau_{d_i}) \tag{11}$$

$$A''(f_i, d_i) = \exp(-\sqrt{(R + j\omega L)(G + jwC)}) \cdot d_i \tag{12}$$

$$g_i'' = \Gamma_{AD} \rho_{2D}^{(i-1)} \Gamma_{2C}^{(i-1)} \tag{13}$$

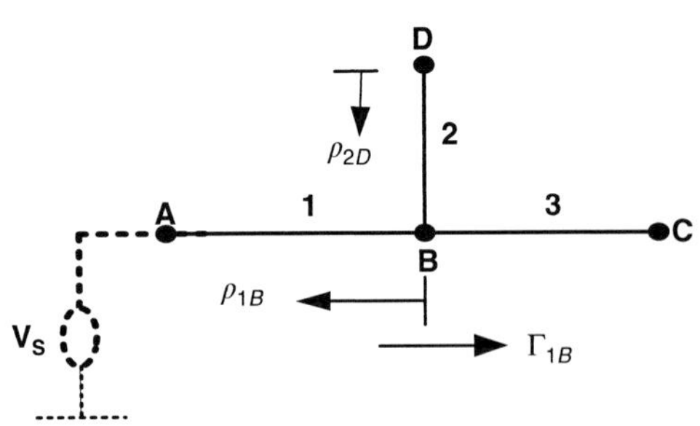

Figure 3: Signal propagation for a PLC with one branch.

4 Anatory *et al.* model

4.1 Power-line network with one interconnection

Consider a TL as shown in Fig. 4; V_s and Z_s are the source voltage and impedance, respectively; AB and BC are TLs with characteristic impedances Z_1 and Z_2, respectively, while Z_{L2} is the load impedance. When the TL is excited with a pulse at point A, signal v^+ will propagate to B. The signal which will propagate through the line is given by $v^+e^{-\gamma_1 l_1}$; where v^+ is terminal voltage, γ_1 is propagation constant and l_1 is arbitrary length that the signal travels towards B. The incident signal wave at point B is given by $v^+e^{-\gamma_1 L_1}$; where L_1 is the length of TL 1. On reaching point B two waves will be generated, the first being the reflected wave back to A and the second wave will travel towards C. The incident wave at C will generate two waves, one will travel towards the load and another one is the reflected wave towards node B. Generally in the first instance there will be three waves traveling between nodes B and C. The first is the direct wave, the second is the wave reflected at node C and the third is the wave reflected at B back to C. The reflections at all points will continue until all signals have been significantly attenuated. In addition the wave reflected at node B from the initial pulse on reaching node A will be reflected back and initiate similar trend. It is perceived that in TL 2 there will be three signals. The generalized expression for this scenario is given by (14) [7–11]. The parameter L is the total number of reflections at point C while M is the number of reflections at the terminal. The parameters ρ_{21}, ρ_S and ρ_{12} are reflection factors from TL 2 to TL 1, that of the source and that of TL 1 to TL 2, respectively. T_{12} and T_{21} are transmission factors from TL 2 to TL 1 and vice versa, respectively. L_2 and l_2 are length of TL 2 and arbitrary length the signal has traveled, respectively. The total received signal at load Z_{L2} is given by (14); where T_{L2} is the transmission factor to load Z_{L2}.

$$V_2 = \sum_{M=1}^{L} T_{L2} a_{21} V_{21} \tag{14}$$

$$a_{21} = \rho_s^{M-1} \rho_{12}^{M-1} e^{-\gamma_1 (2(M-1)L_1)} \tag{15}$$

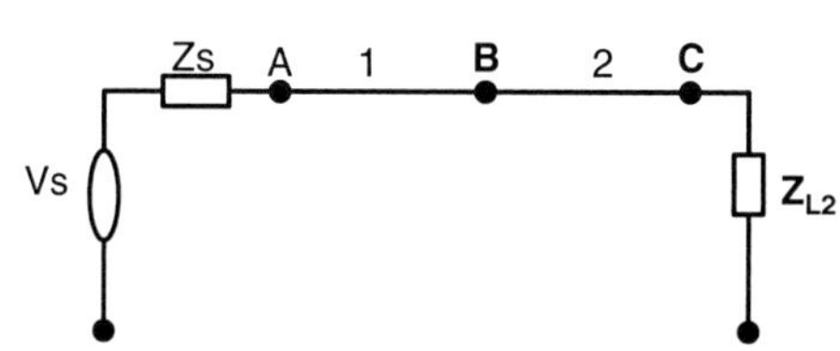

Figure 4: Power-line TL terminated in another line with load connected at the second line.

$$V_{21} = (1+\beta_{21})\beta_2 e^{-\gamma_2 l_2} + \sum_{N=1}^{L} (1+\beta_{21})\beta_2 \rho_{21}^{N} \rho_{L1}^{N} e^{-\gamma_2(2NL_2+l_2)}$$
$$+\sum_{N=1}^{L} (1+\beta_{21})\beta_2 \rho_{21}^{N-1} \rho_{L1}^{N} e^{-\gamma_2(2NL_2-l_2)} \quad (16)$$

$$\beta_{21} = \frac{B_{21}}{1-B_{21}} + \frac{B_{21}A_{21}}{1-B_{21}A_{21}} + \frac{B_{21}A_{21}{}^{2}}{1-B_{21}A_{21}{}^{2}} + \frac{B_{21}A_{21}{}^{3}}{1-A_{21}B_{21}{}^{3}} + .. \quad (17)$$

$$B_{21} = (e^{-\gamma_1 2L_1} T_{12} e^{-\gamma_2 2L_2}) \rho_{L2} T_{21} \rho_s, \quad A_{21} = \rho_{L2} \rho_{21} e^{-\gamma_2 2L_2}, \qquad \beta_2 = T_{12} e^{-\gamma_1 L_1} v^{+} \quad (18)$$

4.2 Power line with one branch at a node

Figure 5 shows a TL with one branch, where L_1, L_2 and L_3 are the lengths of TLs and l_1, l_2 and l_3 are arbitrary distances the signal has traveled from node A to node B, node B to node D and node B to node C, respectively. Z_{L2} and Z_{L3} are the load impedances. Assuming that the loads are not terminated in their characteristic impedances, the received signal at load 2 (Z_{L2}) is the contribution from TL 1 and TL 3. The signals at load 2 from TL 1 (V_{21}) and TL 3 (V_{23}) can be represented by (19). The parameter a_{23} is as shown in (20) whereby each parameter of TL 1 in (19) is replaced by that of TL 3. The other parameters are as given in (22)–(25).

$$V_2 = \sum_{M=1}^{L} (a_1 V_{21} + a_{23} V_{23}) \quad (19)$$

$$a_{23} = \rho_{L3}^{M-1} \rho_{32}^{M-1} e^{-\gamma_3(2(M-1)L_3)} \quad (20)$$

$$V_{23} = (1+\beta_{23})\beta_{132} e^{-\gamma_2 l_2} + \sum_{N=1}^{L} (1+\beta_{23})\beta_{132} \rho_{23}^{N-1} \rho_{L_1}^{N} e^{-\gamma_2(2NL_2-l_2)}$$
$$+\sum_{N=1}^{L} (1+\beta_{23})\beta_{132} \rho_{23}^{N} \rho_{L_1}^{N} e^{-\gamma_2(2NL_2+l_2)} \quad (21)$$

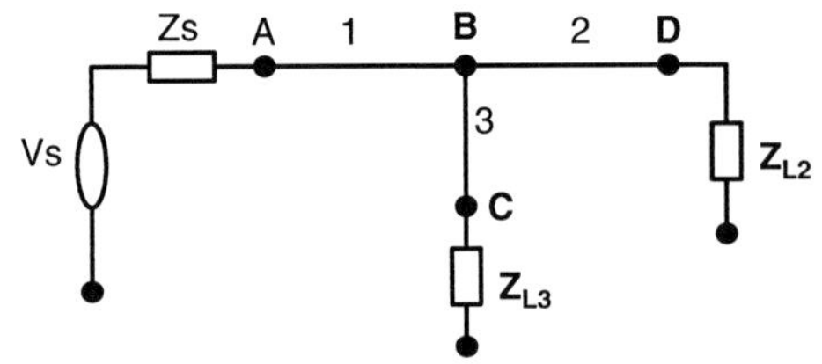

Figure 5: Transmission line with one interconnection and one branch.

$$\beta_{23} = \frac{B_{23}}{1-B_{23}} + \frac{B_{23}A_{23}}{1-B_{23}A_{23}} + \frac{B_{23}A_{23}^{\ 2}}{1-B_{23}A_{23}^{\ 2}} + \frac{B_{23}A_{23}^{\ 3}}{1-A_{23}B_{23}^{\ 3}} + \dots \tag{22}$$

$$\beta_{132} = T_{13}T_{23}\rho_{L_3}e^{-\gamma_1 L_1}e^{-\gamma_3(2L_3)}V_{\mathrm{T}} \tag{23}$$

$$B_{23} = \rho_{L_3}\rho_{L_2}T_{23}T_{32}e^{-\gamma_3(2L_3)}e^{-\gamma_1(2L_2)} \tag{24}$$

$$A_{23} = \rho_{L_3}\rho_{23}e^{-\gamma_2(2L_2)} \tag{25}$$

4.3 Power line with branches distributed at a node

For a TL with multiple branches at a single node (e.g. node B in Fig. 6) the generalized transfer function can be represented by (26a). In (26a), N_{t} is the total number of branches connected at node B and terminated in any arbitrary load. Let n, m, M, H_{mn} (f) and $T_{\mathrm{L}m}$ represent any branch number, any referenced (terminated) load, number of reflections (with total L number of reflections), transfer function between line n and *a* referenced load m, transmission factor at the referenced load m, respectively. With these the signal contribution factor a_{mn} is given by (26b), where ρ_{mn} is the reflection factor at node B between line n and the referenced load m, γ_n is the propagation constant of line $\boldsymbol{n}$ that has line length ln. All terminal reflection factors PLn in general are given by (27c), except at source where $\rho_{\mathrm{L}1} = \rho_{\mathrm{S}}$ is the source reflection factor [7–11]. Also Z_{S} is the source impedance, Z_{n} is the characteristic impedance of any terminal with source while V_{S} and Z_{L} are source voltage and load impedance, respectively, based on Fig. 6.

$$H_m(f) = \sum_{M=1}^{L}\sum_{n=1}^{N_t} T_{\mathrm{L}m}a_{mn}H_{mn}(f) \quad \mathbf{n \neq m} \tag{26a}$$

$$a_{mn} = P_{\mathrm{L}n}^{M-1}\rho_{nm}^{M-1}e^{-\gamma_n(2(M-1)\ell_n)} \tag{26b}$$

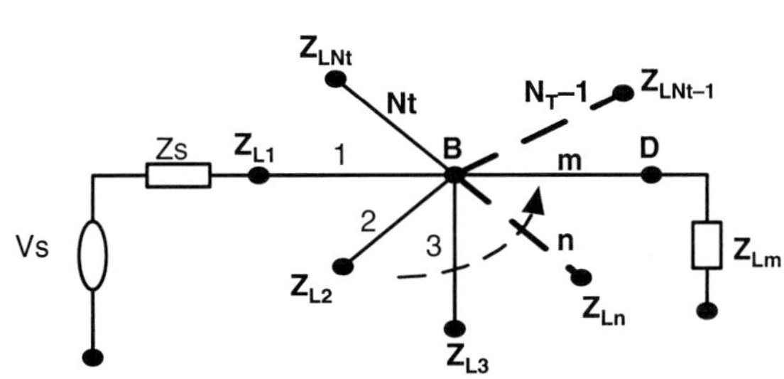

Figure 6: Power-line network with multiple branches at a single node.

$$P_{Ln} = \begin{cases} \rho_s & n = 1(\text{source}) \\ \rho_{Ln}, & \text{otherwise} \end{cases} \tag{26c}$$

The output referenced voltage $V_m(f)$ across any load in frequency domain is given by (27).

$$V_m(f) - H_m(f)\left(\frac{Z_{Ln}}{Z_{Ln} + Z_s}\right)V_s \tag{27}$$

4.4 Power-line network with distributed branches

For a more generalized case applicable to any line configuration consider a power-line network with distributed branches as shown in Fig. 7, the transfer function of such network is given by (28a). In (28a) the parameters used has the same meaning as used above and M_T is the total number of distributed nodes, d is any referenced node (1… M_T), Hmnd (f) is the transfer function from line n to a referenced load m at a referenced node d. All parameters used in (28a–c) are similar to (26a–c), respectively, but with reference node d (Fig. 7).

$$H_{mM_T}(f) = \prod_{d=1}^{M_T}\sum_{M=1}^{L}\sum_{n=1}^{N_T} \mathrm{T}_{Lmd}\alpha_{mnd}H_{mnd}(f) \quad \mathbf{n \neq m} \tag{28a}$$

$$\alpha_{mnd} = P_{Lnd}^{M-1}\rho_{nmd}^{M-1}e^{-\gamma_{nd}(2(M-1)\ell_{nd})} \tag{28b}$$

$$P_{Lnd} = \begin{cases} \rho_s & d = n = 1(\text{source}) \\ \rho_{Lnd}, & \text{otherwise} \end{cases} \tag{28c}$$

$$V_{mM_T}(f) = H_{mM_T}(f)\cdot\left(\frac{Z_{Ldn}}{Z_{Ldn} + Z_s}\right)V_s \tag{29}$$

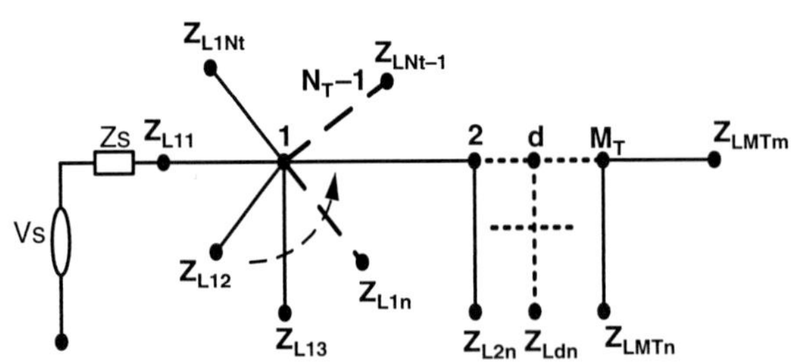

Figure 7: Power-line network with distributed branches.

5 Anatory *et al.* channel model based on generalized TL theory (generalized TL theory model)

Having realized that for obtaining accurate frequency response, for any TL problem Anatory *et al.* [12] used the TL theory and corresponding modal analyses for solutions of currents and voltage along any point on the line to develop more accurate channel model (will be demonstrated when comparisons are made later). In the previous chapter we have seen that for any TL system the current $I(x)$ and voltage $V(x)$ at arbitrary line length x are given by (30) and (31), respectively [12]. In (30) and (31), Z_C, γ, I_m^+ and I_m^- are line characteristic impedance, propagation constant and model currents representing the forward and backward waves, respectively.

$$I(x) = \left(e^{-\gamma x} I_m^+(x) - e^{\gamma x} I_m^-(x)\right) \tag{30}$$

$$V(x) = Z_C \left(e^{-\gamma x} I_m^+(x) + e^{\gamma x} I_m^-(x)\right) \tag{31}$$

In all the transfer function derivations to follow we use the above general equations.

5.1 Power-line network with one branch

The procedure for deriving the transfer function is explained by considering the example of a power-line network with one branch as shown in Fig. 8. In Fig. 8, V_S, Z_s, Z_{C1}, Z_{C2}, Z_{C3}, Z_2 and Z_3 being the source voltage, source impedance, and characteristic impedance of line 1, characteristic impedance of line 2, characteristic impedance of line 3, load impedance of line 2 and load impedance terminated on line 3, respectively. L_1, L_2 and L_3 are the line lengths as shown in Fig. 8. Note that L_2 and L_3 include L_1. In Fig. 8, at point A, $x = 0$, using (30) and (31), the voltage and currents at that particular point is given by (32) and (33), respectively. Voltage V_1 (0) at point A is given by (34) and substituting the respective values

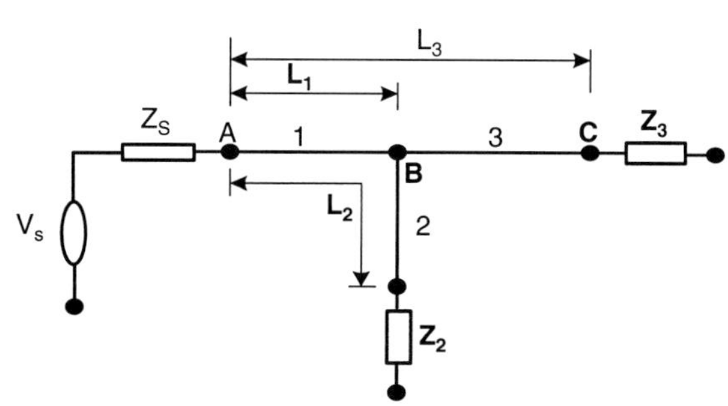

Figure 8: Power-line network with one branch.

from (33) and (34) the results are as in (35). Let $A_1^+ = Z_{C1} + Z_S$ and $A_1^- = Z_{C1} - Z_S$, then (35) can be represented by (36).

$$V_1(0) = Z_{C1}\left(I_{m1}^+ + I_{m1}^-\right) \tag{32}$$

$$I_1(0) = \left(I_{m1}^+ - I_{m1}^-\right) \tag{33}$$

$$V_1(0) = V_s - Z_s I_1(0) \tag{34}$$

From (33)–(35), we get (36) below.

$$\left(Z_{C1} + Z_S\right)I_{m1}^+ + \left(Z_{C1} - Z_S\right)I_{m1}^- = V_S \tag{35}$$

Defining $Z_{C1} + Z_S = A_1^+$, and $Z_{C1} - Z_S = A_1^-$ in (36), we get,

$$A_1^+ I_{m1}^+ + A_1^- I_{m1}^- = V_S \tag{36}$$

Consider point B at $x = L_1$, let γ_1, γ_2 and γ_3 be the propagation constants for line 1, line 2 and line 3, respectively. At B, the voltage is continuous $V_1(L_1) = V_2(L_1) = V_3(L_1)$ and the current is discontinuous, $I_1(L_1) = I_2(L_1) + I_3(L_1)$. Using in (23) and (24), $V_1(L_1) = Z_{C1}(e^{-\gamma_1 L_1} I_{m1}^+ + e^{\gamma_1 L_1} I_{m1}^-)$, $V_2(L_1) = Z_{C2}(e^{-\gamma_2 L_1} I_{m2}{}^+ + e^{\gamma_2 L_1} I^-{}_{m2})$ and $V_3(L_1) = Z_{C3}(e^{-\gamma_3 L_1} I_{m3}{}^+ + e^{\gamma_2 L_1} I^-{}_{m3})$ and for currents $I_1(L_1) = (e^{-\gamma_1 L_1} I_{m1}^+ - e^{\gamma_1 L_1} I_{m1}^+)$, $I_2(L_1) = (e^{-\gamma_2 L_1} I_{m2}^+ - e^{\gamma_2 L_1} I_{m2}^-)$ and $I_3(L_1) = (e^{-\gamma_3 L_1} I_{m3}^+ - e^{\gamma_3 L_1} I_{m3}^-)$.

Consider now at point E, $x = L_2$ and point C, $x = L_3$, the current in the loads are $I_2(L_2) = V_2(L_2)/Z_2$ and $I_3(L_3) = V_3(L_3)/Z_3$, where, using (30) and (31) $I_2(L_2) = (e^{-\gamma_2 L_2} I_{m2}^+ - e^{\gamma_2 L_2} I_{m2}^-)$, $I_3(L_3) = (e^{-\gamma_3 L_3} I_{m3}^+ - e^{\gamma_3 L_3} I_{m3}^-)$, $V_2(L_2) = Z_{C2}(e^{-\gamma_2 L_2} I_{m2}^+ + e^{\gamma_2 L_2} I_{m2}^-)$ and $V_3(L_3) = Z_{C3}(e^{-\gamma_3 L_3} I_{m3}^+ + e^{\gamma_3 L_3} I_{m3}^-)$. Solving the linear simultaneous expressions so obtained at all nodes the values for I_{m1}^+, I_{m1}^-, I_{m2}^+, I_{m2}^-, I_{m3}^+ and I_{m3}^- can be obtained for any frequency. The expression for transfer functions $H(f)$ relating the voltage at node C to node A can be obtained.

5.2 Branches concentrated at one node

Consider the configuration as given in Fig. 9, with the line and load parameters as discussed in previous section. Using the procedure discussed in the previous section for writing the boundary condition expressions at every node the generalized transfer function $H_m(f)$ between any load termination and the sending end is given by (37a). In general the parameters Z_{Cm}, Z_m, γ_m and L_m are characteristic impedance for line m, load impedance for line m, propagation constant for line m and length for line which is the shortest distance measured from point A to any point m at the load, respectively.

$$H_m(f) = \frac{Z_{C1} + Z_s}{Z_{C1}} Z_{Cm}\left(e^{-\gamma_m L_m}\beta_m + e^{\gamma_m L_m}\right)A_{1m}\frac{1}{A_1^+\beta_1 + A_1^-} \tag{37a}$$

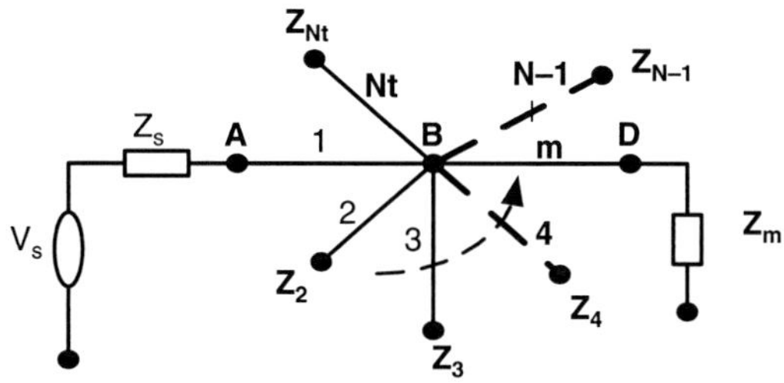

Figure 9: Power line with branches concentrated at one node.

$$A_{1m} = \frac{a_1}{a_{1m}} \tag{37b}$$

$$\beta_m = \frac{(1 - Z_{Cm}/Z_m)e^{-\gamma_m L_m}}{(1 + Z_{Cm}/Z_m)e^{\gamma_m L_m}} \quad m = 2,3 \quad N_t \tag{37c}$$

$$\beta_1 = \frac{C_1^- - P_1 B_1^-}{C_1^+ + P_1 B_1^+} \tag{37d}$$

$$P_1 = \frac{e_{12}}{a_{12}} + \frac{e_{13}}{a_{13}} + ... + \frac{e_{1m}}{a_{1m}} \tag{37e}$$

$$e_{1m} = C_{m1}^- - C_{m1}^+ \beta_m \tag{37f}$$

$$a_{1m} = B_{m1}^+ \beta_m + B_{m1}^- \tag{37g}$$

$$a_1 = B_1^+ \beta_1 + B_1^- \tag{37h}$$

$$B_{m1}^+ = Z_{Cm} e^{-\gamma_m L_1} \tag{37i}$$

$$B_{m1}^- = Z_{Cm} e^{\gamma_m L_1} \tag{37j}$$

$$C_{m1}^+ = e^{-\gamma_m L_1} \tag{37k}$$

$$C_{m1}^- = e^{\gamma_m L_1} \tag{37l}$$

In the above equations $B_1^+ = Z_{C1}e^{-\gamma_1 L_1}$, $B_1^- = Z_{C1}e^{\gamma_1 L_1}$, $C_1^+ = e^{-\gamma_1 L_1}$ $C_1^- = e^{\gamma_1 L_1}$, $A_1^+ = Z_{C1} + Z_S$ and $A_1^- = Z_{C1} - Z_S$. In 37(c) N_t is the total number of branches connected at node B.

5.3 Distributed branches along the line section

Consider the power-line network as shown in Fig. 10. Note that the procedure for obtaining the transfer function is the same, i.e. writing the voltage and current boundary conditions at all the nodes and solving for the unknown modal currents. The transfer function for the voltage between any load point Z_{nm} and the sending end is given by (38a). In (38a) Z_{Cnm}, Z_{nm}, γ_{nm}, L_{nm} and L_n are characteristic impedance of line segment *nm*, terminal load impedance of line *nm*, propagation constant of line segment *nm*, shortest length of line segment *nm* and shortest line length from the sending end to the node *n* under consideration, respectively. Note that all parameters with *nn* means the consideration is at the node.

$$H_{nm}(f) = \frac{Z_{C11} + Z_s}{Z_{C11}} Z_{Cnm}\left(e^{-\gamma_{nm}L_{nm}}\beta_{nm} + e^{\gamma_{nm}L_{nm}}\right)A_{nm}\frac{1}{A_1^+\beta_{11} + A_1^-} \quad \text{(38a)}$$

$$\beta_{nn} = \frac{C_{nn}^- - P_n B_{nn}^-}{C_{nn}^+ + P_n B_{nn}^+} \quad \text{(38b)}$$

$$P_n = \frac{e_{nm(1)}}{a_{nm(1)}} + \frac{e_{nm(2)}}{a_{nm(2)}} \quad \text{(38c)}$$

$$\beta_{nm} = \frac{(1 - Z_{Cnm}/Z_{nm})e^{-\gamma_{nm}L_{nm}}}{(1 + Z_{Cnm}/Z_{nm})e^{\gamma_{nm}L_{nm}}} \quad \text{(38d)}$$

$$A_{nm} = \frac{a_n a_{n-1} \cdots\cdots\cdots a_1}{a_{nm}a_{(n-1)(n)}a_{(n-2)(n-1)} \cdots\cdots\cdots\cdots a_{12}} \quad \text{(38e)}$$

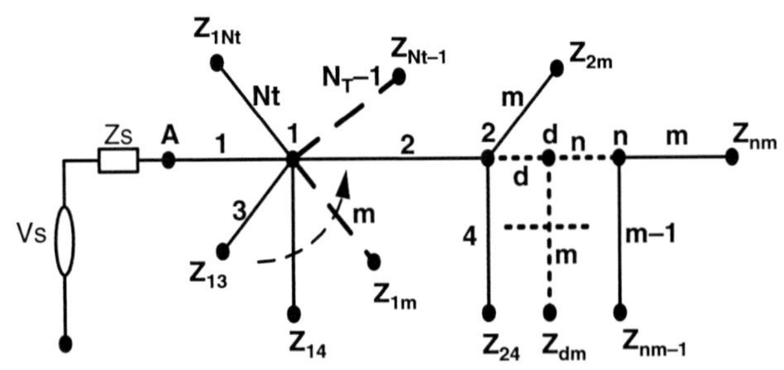

Figure 10: Power-line network with distributed branches.

$$a_{nm} = B_{mn}^{+}\beta_{n} + B_{mn}^{-} \tag{38f}$$

$$e_{nm} = C_{mn}^{-} - C_{mn}^{+}\beta_{n} \tag{38g}$$

$$\beta_{n} = \begin{cases} \beta_{(n+1)(n+1)} & \text{to node} \quad n+1 \\ \beta_{nm} & \text{to load} \quad m \end{cases} \tag{38h}$$

$$a_{mn} = B_{nm}^{+}\beta_{nm} + B_{nm}^{-} \tag{38i}$$

$$a_{n} = B_{nn}^{+}\beta_{nn} + B_{nn}^{-} \tag{38j}$$

$$B_{mn}^{+} = Z_{\mathrm{C}mn}e^{-\gamma_{mn}L_{n}} \tag{38k}$$

$$B_{mn}^{-} = Z_{\mathrm{C}mn}e^{\gamma_{mn}L_{n}} \tag{38l}$$

$$B_{nm}^{+} = Z_{\mathrm{C}nm}e^{-\gamma_{nm}L_{n}} \tag{38m}$$

$$B_{nm}^{-} = Z_{\mathrm{C}nm}e^{\gamma_{nm}L_{n}} \tag{38n}$$

$$C_{mn}^{+} = e^{-\gamma_{mn}L_{n}} \tag{38o}$$

$$C_{mn}^{-} = e^{\gamma_{mn}L_{n}} \tag{38p}$$

$$B_{nn}^{+} = Z_{\mathrm{C}nn}e^{-\gamma_{nn}L_{n}} \tag{38q}$$

$$B_{nn}^{-} = Z_{\mathrm{C}nn}e^{\gamma_{nn}L_{n}} \tag{38r}$$

$$C_{nn}^{+} = e^{-\gamma_{nn}L_{n}} \tag{38s}$$

$$C_{nn}^{-} = e^{\gamma_{nn}L_{n}} \tag{38t}$$

In (38a), $A_{1}^{+} = Z_{\mathrm{C}11} + Z_{\mathrm{S}}$ and $A_{1}^{-} = Z_{\mathrm{C}11} - Z_{\mathrm{S}}$.

6 The validity of Zimmermann and Dostert and its improvements

It has been identified by Anatory *et al.* [15] that the Zimmermann and Dostert model [5] deviates in terms of delay compared to ATP-EMTP which is implemented based on TL theory. For example, consider the power-line network shown in Fig. 11, where Z_S, V_s, Z_{L1} and Z_{L2} are source impedance, source voltage, load impedance at node C and load impedance at node D, respectively. The lengths of line segments AB, BD and BC were considered as 60, 200 and 100 m, respectively. Per unit length inductances and capacitances were taken as 0.44388 μH/m and 0.61734 pF/m, respectively, for all the line segments. A 2-V rectangular pulse with width of 1 μs shifted by 0.5 μs was considered as the voltage source injection. The Zimmermann and Dostert [5] model with 10 paths was considered. In the simulations Z_{L1} was kept at 20 Ω while Z_S and Z_{L2} were terminated in the characteristic impedance. The voltage was calculated across Z_{L2}. Figure 12a shows

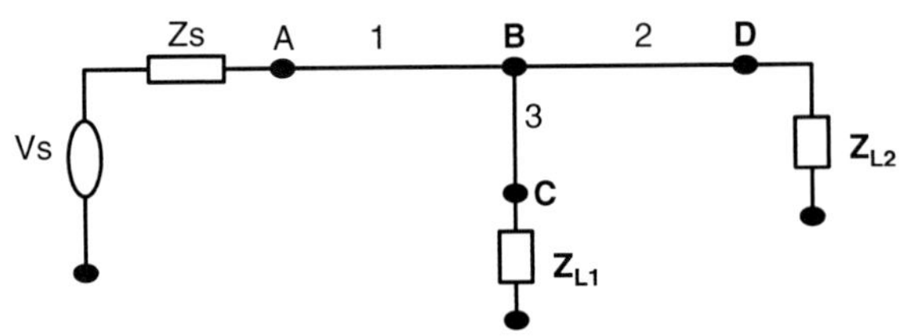

Figure 11: Case 1: power-line network configuration with one branch.

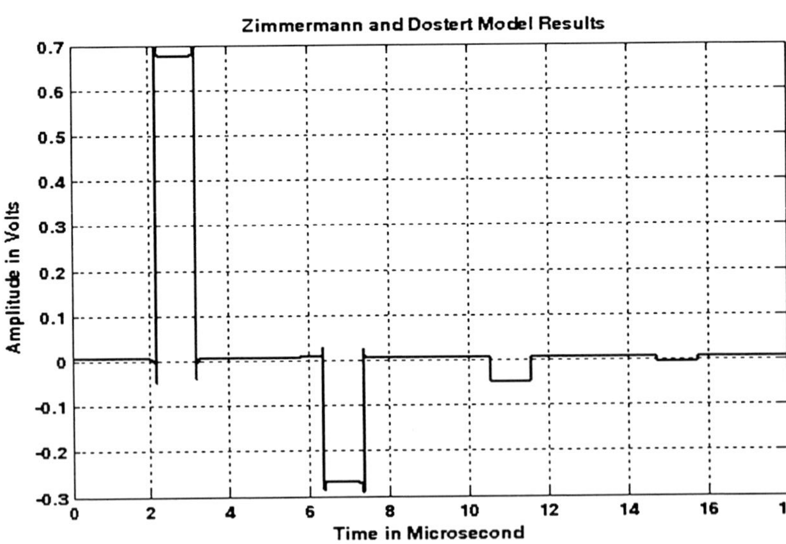

Figure 12a: Zimmermann and Dostert model results for PLC models for a power-line network with one branch.

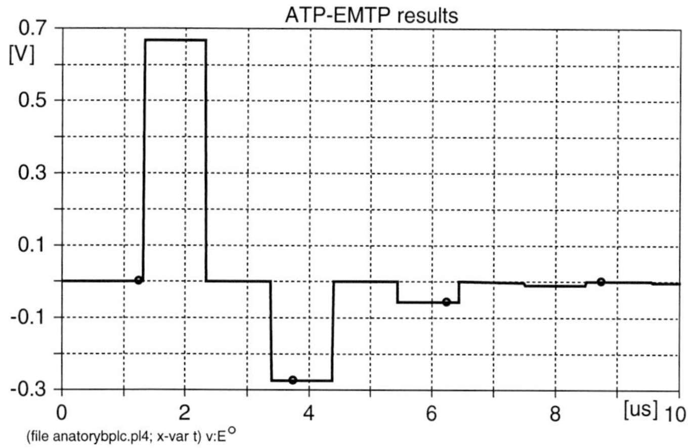

Figure 12b: Simulations using ATP-EMTP software for a power-line network with one branch.

the simulation results for the Zimmermann and Dostert model. For checking the validity of the models, the configuration was implemented in ATP-EMTP software and the corresponding results are shown in Fig. 12b. It is found that the amplitude for both the Zimmermann and Dostert model and that predicted by ATP-EMTP are similar, while the time delay in the two models are different.

Let us look at the way in which the delay for the Zimmermann and Dostert model can be reduced. Looking at the model this can be attributed to the distance parameter in the attenuation factors in the model. Removing the attenuation factor in the model, it can be implemented as (39). Deriving the parameter from the actual network, the expression can be implemented as (40). The same network as before was solved with the modified Zimmermann and Dostert model using (40). Figure 13 shows the simulation results for the modified Zimmermann and Dostert model. It can be observed that removing the distance parameter in the attenuation factor gives better results. In the following sections, the simulations will be based on the modified Zimmermann and Dostert models.

$$H(f) = \sum_{i=1}^{N} g_i e^{-(a_0 + a_1 f^k)} e^{-j2\pi f \frac{d_i}{v_p}} \tag{39}$$

$$H(f) = \sum_{i=1}^{N} g_i''(f) A''(f_i) \cdot \exp\left(-j2\pi f \frac{d_i}{v_p} \right) \tag{40a}$$

$$A''(f_i) = \exp(-\sqrt{(R + j\omega L_e)(G + jwC_e)}) \tag{40b}$$

$$g_i'' = \Gamma_{AD} \rho_{2D}^{(i-1)} \Gamma_{2C}^{(i-1)} \tag{40c}$$

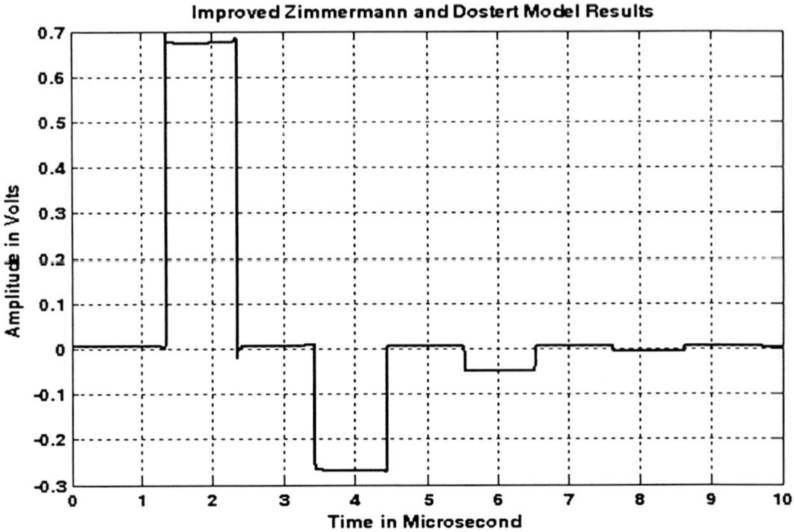

Figure 13: Improved Zimmermann and Dostert Model results for PLC models for a power-line network with one branch.

7 Comparison between different channel models – case studies

7.1 Case 1: power-line network with one branch

Consider the power-line network shown in Fig. 11, where Z_S, V_s, Z_{L1} and Z_{L2} are source impedance, source voltage, load impedance at node C and load impedance at node D, respectively. The lengths of line segments AB, BD and BC were considered as 60, 200 and 100 m, respectively. Per unit length inductances and capacitances of line segments AB, BD and BC were taken as 0.44388 μH/m and 0.61734 pF/m, respectively. A 2-V rectangular pulse with width of 1 μs shifted by 0.5 μs was considered as the voltage source injection. Philipps [3], modified Zimmermann and Dostert, Anatory et al. [7] and generalized TL theory model [12] were applied for comparisons. In the cases of Philipps, modified Zimmermann and Dostert models 10 paths were considered. In Anatory *et al.* [7] model 10 total numbers of reflections were considered. In the simulations Z_{L1} was kept open while Z_S and Z_{L2} were terminated in the characteristic impedance. The voltage was calculated across Z_{L2}. Figure 14a shows the simulation results for Philipps, modified Zimmermann and Dostert, Anatory *et al.* and generalized TL theory models. It is observed that all four models have similar results. The validity of the models for this configuration was implemented in ATP-EMTP software and the corresponding results are shown in Fig. 14b, which confirms that all four models are consistent.

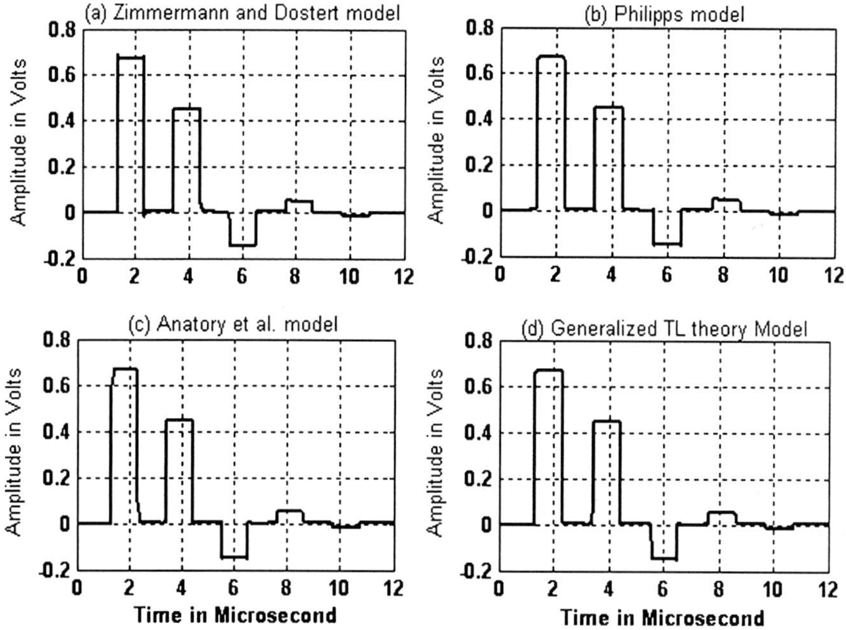

Figure 14a: Comparisons for PLC models for a power-line network with one branch as shown in Fig. 11 for Case 1, voltage response calculated at Z_{L2}.

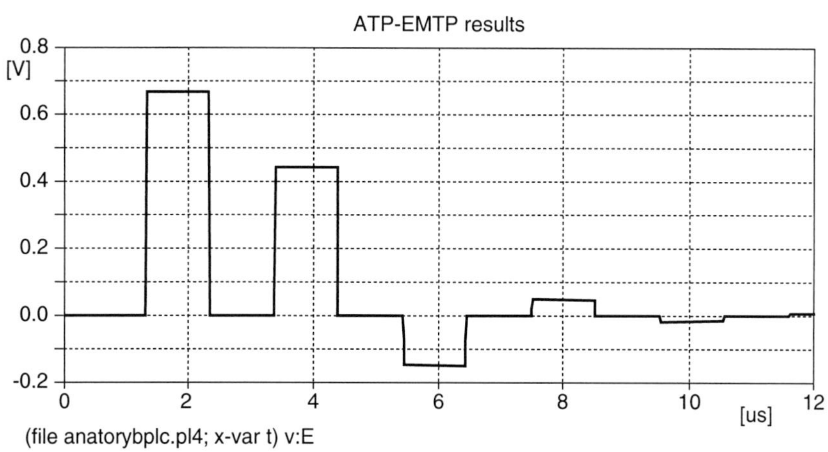

Figure 14b: Simulations using ATP-EMTP software for a power-line network with one branch as shown in Fig. 11 for Case 1, voltage response calculated at Z_{L2}.

7.2 Case 2: power-line network with two branches at the same node

Consider the power-line network shown in Fig. 15, where Z_S, V_s, Z_{L2}, Z_{L3} and Z_{L4} are source impedance, source voltage, load impedance at nodes C, D and E, respectively. The lengths of line segments AB, BD, BC and BE were considered as 60, 200, 100 m and 100 m, respectively. Per unit length inductances and capacitances of line segments AB, BD, BC and BE were taken as 0.44388 μH/m and 0.61734 pF/m, respectively. A 2-V rectangular pulse with width of 1 μs shifted by 0.5 μs was considered. Philipps [3], modified Zimmermann and Dostert, Anatory *et al.* [7] and generalized TL method [12] were applied. In the cases of Philipps, modified Zimmermann and Dostert models 10 paths were considered. In Anatory *et al.* [7] model 10 total numbers of reflections were considered. In the investigation Z_{L4} was kept open while Z_S, Z_{L2} and Z_{L3} were terminated in the characteristic impedance. The voltage was calculated across Z_{L3}. Figure 16a shows the simulation results for Philipps, modified Zimmermann and Dostert models, Anatory *et al.* model and generalized TL theory model. It is observed that all Philipps, modified Zimmermann and Dostert models, Anatory *et al.* models have similar results. But the generalized TL theory model shows a different response. The same configuration was implemented in ATP-EMTP software and the results are shown in Fig. 16b.

As expected the generalized TL theory model being more accurate is consistent with the corresponding ATP-EMTP simulations. This also demonstrates that the generalized TL theory channel model is the one that needs to be used for channel performance analysis.

7.3 Case 3: power-line network with two distributed branches along the line between sending and receiving ends

Consider the power-line network shown in Fig. 17, wherein Z_S, V_s, Z_{L1}, Z_{L2} and Z_{L3} are source impedance, source voltage, load impedance at nodes E, D and F, respectively. The length of line segments AB, BC, BE, CD and CF are 200 m each.

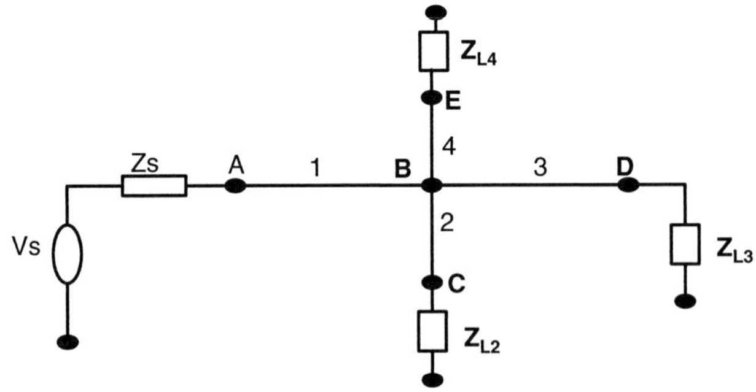

Figure 15: Case 2: power-line network configuration with two branches at the same node.

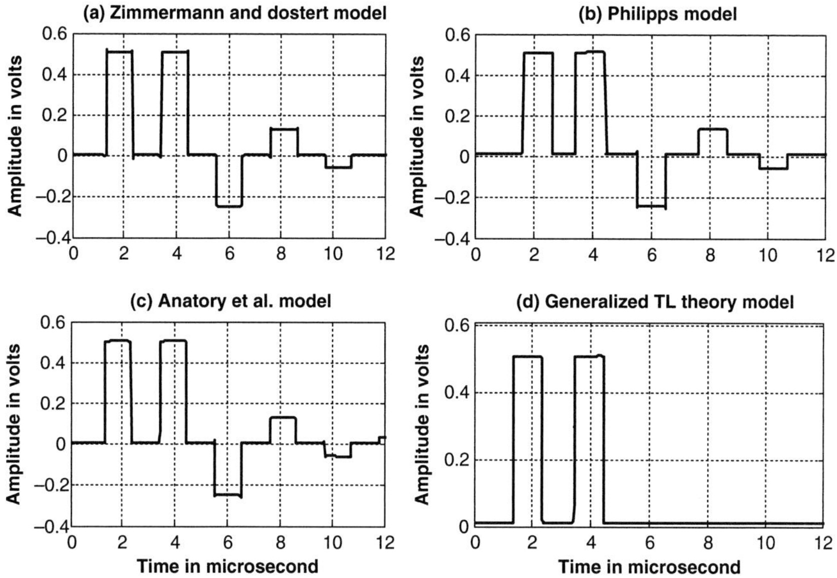

Figure 16a: Comparisons for PLC models for a power-line network with two branches at a node as shown in Fig. 15 for Case 2, voltage response calculated at Z_{L3}.

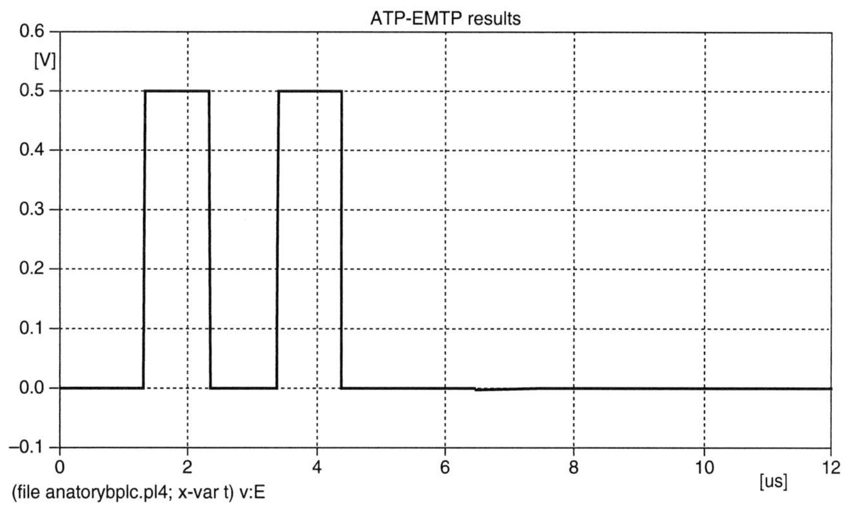

Figure 16b: Simulations using ATP-EMTP software for a power-line network with two branches at a node as shown in Fig. 15 for Case 2, voltage response calculated at Z_{L3}.

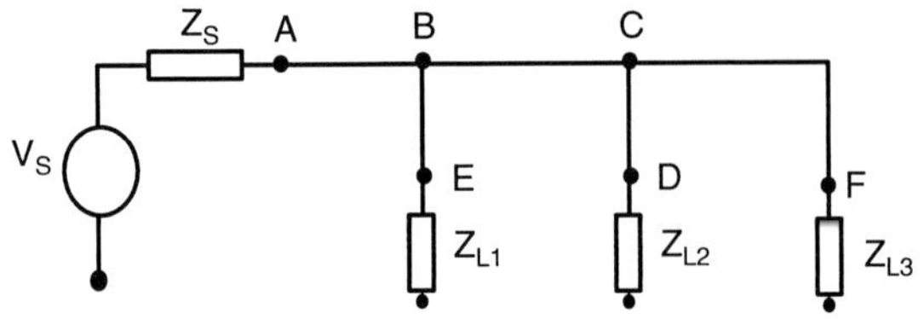

Figure 17: Case 3: network configuration with distributed branches.

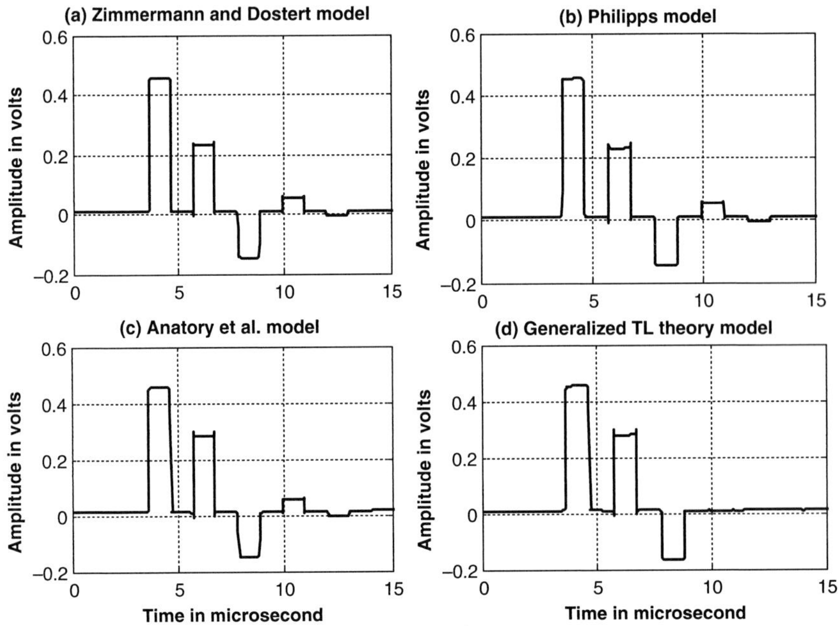

Figure 18a: Comparisons for PLC models for a power-line network with two distributed branches as shown in Fig. 17 for Case 3, voltage response calculated at Z_{L3}.

Per unit length inductances and capacitances of all line segments are taken as 0.44388 μH/m and 0.61734 pF/m, respectively. A 2-V rectangular pulse with width of 1 μs shifted by 0.5 μs was considered as the voltage source injection. Philipps [3], modified Zimmermann and Dostert, Anatory *et al.* [7] and generalized TL theory model [12] were applied. In the cases of Philipps, modified Zimmermann and Dostert models 10 paths were considered. In Anatory *et al.* [7] model 10 total numbers of reflections were considered. In the investigation Z_{L1} and Z_{L2} were kept at 50 Ω and open, respectively, while Z_S and Z_{L3} were terminated at 85 Ω.

The voltage response was calculated at Z_{L3}. Figure 18a shows the simulation results for Philipps, modified Zimmermann and Dostert models, Anatory *et al.* model and generalized TL theory model. It is observed that Philipps,

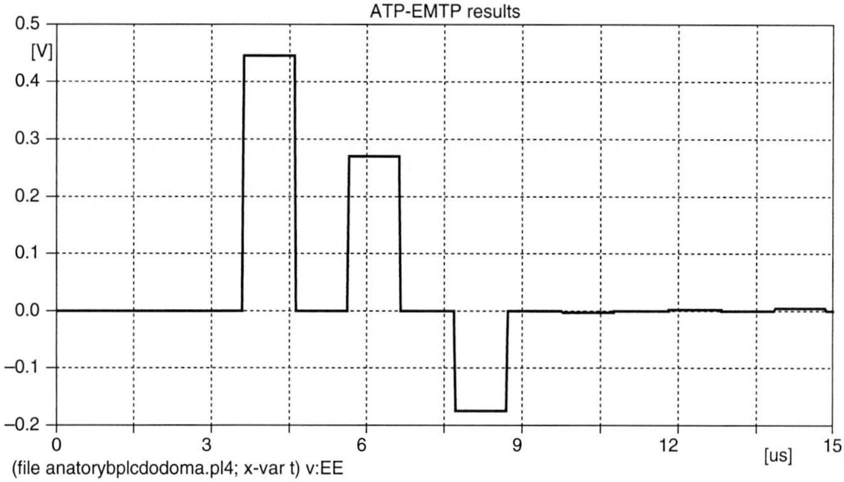

Figure 18b: Simulations using ATP-EMTP software for a power-line network with two distributed branches as shown in Fig. 17 for Case 3, voltage response calculated at Z_{L3}.

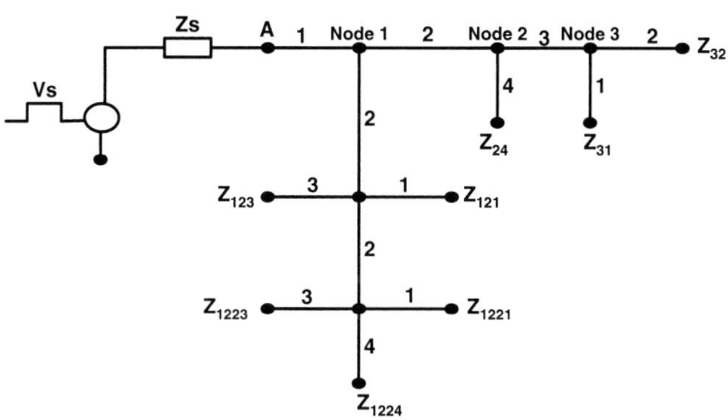

Figure 19: Power-line configuration with tree structure.

modified Zimmermann and Dostert, Anatory *et al.* models have similar results while generalized TL theory model predicts different responses after about 10 μs. The configuration was implemented in ATP-EMTP software, and the corresponding simulations are shown in Fig. 18b. As expected again the generalized TL theory model predictions are consistent with the ATP-EMTP result, which indicates the accuracy of the generalized TL theory model compared to other models.

7.4 Case 4: power line with tree structure

To evaluate the strength of the generalized TL theory channel model as applicable to any TL network topology, consider the tree configuration as shown in Fig. 19.

All the TL segments are 500 m. The source impedance Z_s and the receiving end Z_{32} were terminated in 456 Ω, while other loads are terminated in 50 Ω. All lines have per unit length inductance and capacitance L_e = 1.64 μH/m and C_e = 7.84 pF/m. The configuration was excited by the same rectangular pulse voltage as discussed earlier. The voltage at Z_{32} based on the proposed channel model derivation is shown in Fig. 20a. The same case was simulated using the ATP-EMTP software and the result is in shown in Fig. 20b. Again it is seen that results are comparable.

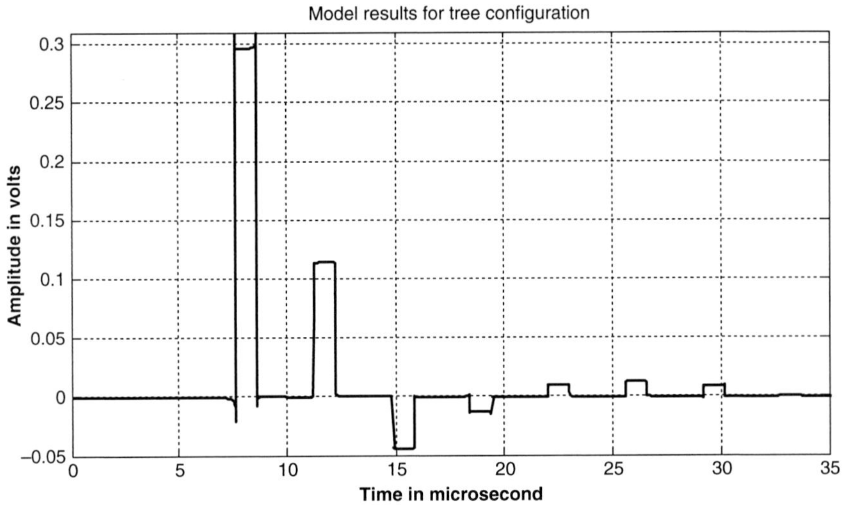

Figure 20a: Voltage response at Z_{32} based on generalized TL theory channel model for a power-line network shown in Fig. 19 corresponding to Case 4.

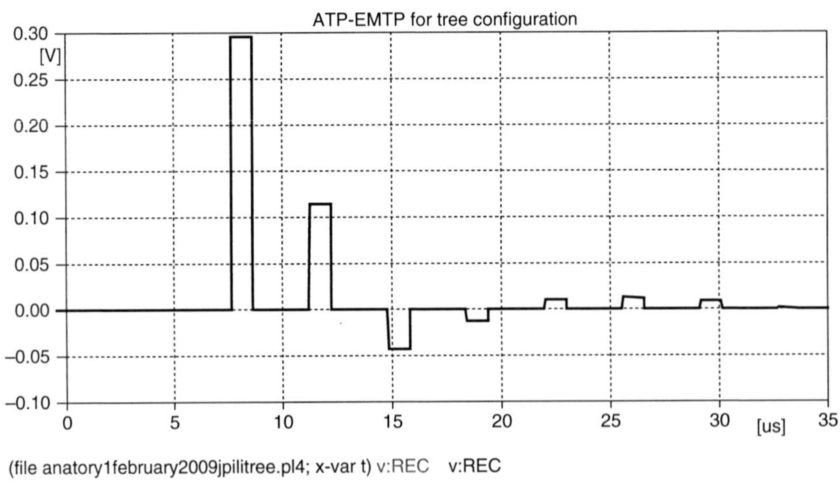

Figure 20b: Voltage response at Z_{32} based on ATP-EMTP software for a power-line network as shown in Fig. 19 corresponding to Case 4.

8 Network transfer functions for coupled TL branches – multiconductor case

The generalized TL theory channel model is not extended to the case of coupled multiconductor TL system, wherein L_e and C_e, the inductance (impedance) and admittance (capacitance) matrices as discussed in the previous chapter are symmetric decided by the number of lines and the line geometry [14]. The current vector $I(x)$ and voltage vector $V(x)$ at arbitrary line length x are given by (41) and (42), respectively [13]. We use for derivations here the same principles and methods as discussed in the earlier section for generalized TL theory channel model derivation. In (41) and (42), T_y, Z_{Cy}, γ_y, I_{my}^{+} and I_{my}^{-} are the transformation matrix, characteristic impedance matrix, propagation constant matrix, modal current vector for forward waves and modal current vector for backward waves for given line section numbered y, respectively.

$$I_y(x) = T_y\left(e^{-\gamma_y x} I_{my}^{+}(x) - e^{\gamma_y x} I_{my}^{-}(x)\right) \tag{41}$$

$$V_y(x) = Z_{C_y} T_y\left(e^{-\gamma_y x} I_{my}^{+}(x) + e^{\gamma_y x} I_{my}^{-}(x)\right) \tag{42}$$

$$\gamma_y = \sqrt{T^{-1} YZT} \tag{43}$$

$$Z_C = ZT\gamma^{-1}T^{-1} \tag{44}$$

$$Z = 2\pi f L_e \tag{45}$$

$$Y = 2\pi f C_e \tag{46}$$

8.1 Power-line network with one branch

Consider a multiconductor TL system as shown in Fig. 21. Let V_S, Z_s, Z_{C1}, Z_{C2}, Z_{C3}, Z_2 and Z_3 be the source voltage vector, source impedance matrix, characteristic

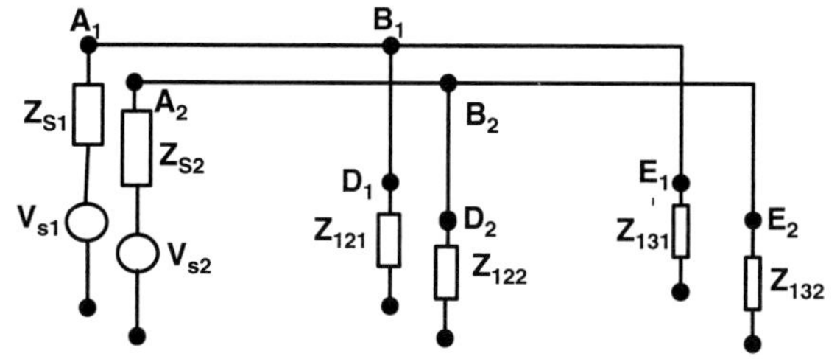

Figure 21: Power line with two branches at the single node.

impedance matrix for a coupled TL segment AB, characteristic impedance matrix for a coupled TL segment BD, characteristic impedance matrix for coupled TL segment BE, terminated load impedance matrix at point D and terminated load impedance matrix at point E, respectively. Also let L_1, L_2, and L_3 be the line lengths from points A to B, A to D and A to E, respectively. Consider Fig. 21, at point A, i.e. at $x = 0$, the voltage and currents [using (41) and (42)] are given by (47) and (48), respectively. As per the boundary condition at point A, the voltage $V_1(0)$ is given by (49). The modal currents in the line section 1 between AB are related to V_S through (50), where, $A_1^+ = (Z_{C1} + Z_S)T_1$ and $A_1^- = (Z_{C1} - Z_S)T_1$.

$$V_1(0) = Z_{C1}T_1\left(I_{m1}^+ + I_{m1}^-\right) \tag{47}$$

$$I_1(0) = T_1\left(I_{m1}^+ - I_{m1}^-\right) \tag{48}$$

$$V_1(0) = V_S - Z_S I_1(0) \tag{49}$$

$$A_1^+ I_{m1}^+ + A_1^- I_{m1}^- = V_S \tag{50}$$

Consider the point B located at length $x = L_1$ from point A, that has two branch sets; let γ_1, γ_2 and γ_3 be the propagation constant matrices for line segments AB, BD and BE, respectively. At B, the voltage is continuous $V_1(L_1) = V_2(L_1) = V_3(L_3)$ and the current is discontinuous, $I_1(L_1) = I_2(L_1) + I_3(L_1)$. Using (1) and (2), the voltage and modal currents in the line sections 1, 2 and 3 are related as $V_1(L_1) = Z_{C1}T_1(e^{-\gamma_1 L_1}I_{m1}^+ + e^{\gamma_1 L_1}I_{m1}^-)$, $V_2(L_1) = Z_{C2}T_2(e^{-\gamma_2 L_1}I_{m2}^+ + e^{\gamma_2 L_1}I_{m2}^-)$ and $V_3(L_1) = Z_{C3}T_3(e^{-\gamma_3 L_1}I_{m3}^+ + e^{\gamma_2 L_1}I_{m3}^-)$ and for currents $I_1(L_1) = T_1(e^{-\gamma_1 L_1}I_{m1}^+ - e^{\gamma_1 L_1}I_{m1}^+)$, $I_2(L_1) = T_2(e^{-\gamma_2 L_1}I_{m2}^+ - e^{\gamma_2 L_1}I_{m2}^-)$ and $I_3(L_1) = T_3(e^{-\gamma_3 L_1}I_{m3}^+ - e^{\gamma_3 L_1}I_{m3}^-)$.

Consider now at point D, $x = L_2$ and point E, $x = L_3$, the current in the loads are $I_2(L_2) = Z_2^{-1}V_2(L_2)$ and $I_3(L_3) = Z_3^{-1}V_3(L_3)$, where, using (41) and (42) the voltage and modal currents in the line sections 2 and 3 are also related as $I_2(L_2) = T_2(e^{-\gamma_2 L_2}I_{m2}^+ - e^{\gamma_2 L_2}I_{m2}^-)$, $I_3(L_3) = T_3(e^{-\gamma_3 L_3}I_{m3}^+ - e^{\gamma_3 L_3}I_{m3}^-)$, $V_2(L_2) = Z_{C2}T_2(e^{-\gamma_2 L_2}I_{m2}^+ + e^{\gamma_2 L_2}I_{m2}^-)$ and $V_3(L_3) = Z_{C3}T_3(e^{-\gamma_3 L_3}I_{m3}^+ + e^{\gamma_3 L_3}I_{m3}^-)$. Solving all the above system of linear simultaneous expressions all the unknown modal currents for all the line sections 1, 2 and 3 can be obtained for any frequency. The expression for transfer functions can, therefore, be derived by relating the voltage at any output node to the sending end node.

8.2 Number of branches concentrated at single node

Consider the configuration as given in Fig. 22, where there are m numbers of branch sets connected to the node B. The generalized transfer function matrix H_m (*f*) between any load termination m and the sending end is given by (51a). In (51), the parameters Z_S, Z_{C1}, Z_{Cm}, Z_m, γ_1, γ_m, L_1, L_m, T_1, T_m are source impedance matrix, characteristic impedance matrix for line section 1 (between A and B),

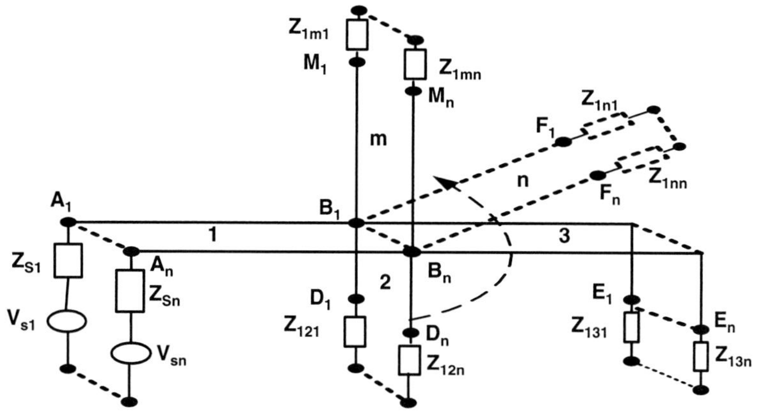

Figure 22: Power line with number of branches at the single node.

characteristic impedance matrix for any other line section m, load impedance matrix m, propagation constant matrix for line section 1, propagation constant matrix for line section m, length for line section 1, length for line mplus length of line section 1, transformation matrix for line section 1 and transformation matrix for line m, respectively.

$$H_m(f) = Z_{C1}^{-1}(Z_{C1} + Z_S)\alpha_m a_{1m}^{-1} a_1 \left(A_1^+ \beta_1 + A_1^-\right)^{-1} \tag{51a}$$

$$\alpha_m = Z_{Cm} T_m \left(e^{-\gamma_m L_m} \beta_m + e^{\gamma_m L_m}\right) \tag{51b}$$

$$\beta_m = \left(T_m + Z_m^{-1} Z_{Cm} e^{+\gamma_m L_m}\right)^{-1} \left(T_m - Z_m^{-1} Z_{Cm} e^{-\gamma_m L_m}\right) \tag{51c}$$

$$\beta_1 = \left(C_1^+ + P_1 B_1^+\right)^{-1} \left(C_1^- - P_1 B_1^-\right) \tag{51d}$$

$$P_1 = e_{12} a_{12}^{-1} + e_{13} a_{13}^{-1} + \ldots\ldots.. + e_{1m} a_{1m}^{-1} \tag{51e}$$

$$e_{1m} = C_{m1}^- - C_{m1}^+ \beta_m \tag{51f}$$

$$a_{1m} = B_{m1}^+ \beta_m + B_{m1}^- \tag{51g}$$

$$B_{m1}^+ = Z_{Cm} T_m e^{-\gamma_m L_1} \tag{51h}$$

$$C_{m1}^+ = T_m e^{-\gamma_m L_1} \tag{51i}$$

$$C_{m1}^{-} = T_m e^{\gamma_m L_1} \tag{51j}$$

In the above equations $B_1^+ = Z_{C1} T_1 e^{-\gamma_1 L_1}$, $B_1^- = Z_{C1} T_1 e^{\gamma_1 L_1}$, $C_1^+ = T_1 e^{-\gamma_1 L_1}$, $C_1^- = T_1 e^{\gamma_1 l_1}$, $A_1^+ = (Z_{C1} + Z_S) T_1$ and $A_1^- = (Z_{C1} - Z_S) T_1$.

8.3 Generalized expression for a network with distributed branches

Consider the power line as shown in Fig. 23. Note that the procedure for obtaining the transfer function is the same, i.e. writing the voltage and current boundary conditions at all nodes and solving for the unknown modal currents. The transfer function H_{nm} can be obtained (52a), which is ratio of the voltage at the load termination of the line m connected to node n to the launched voltage at node A in Fig. 23. In this case it is seen that at each node excluding the source node we have branches concentrated at all possible given nodes 1, 2, 3, … n. The number of branches at a given nodes can vary as 1, 2, ... m. The notation of each parameter is explained as follows. The characteristic impedance Z_{Cmn} is such that, it represents the characteristic impedance of a line m connected to node n. The propagation constant matrix γ_{mn} and the transformation matrix T_{mn} are also defined in the same way. L_{nm} is the length of the line m connected to node n plus the distance (L_n) between the source and node n. The characteristic impedance Z_{Cnm} is such that it represents the characteristic impedance of a line n connected to line m. The propagation constant matrix γ_{nm} and the transformation matrix T_{nm} are also defined in the same way. The characteristic impedance Z_{Cnn} is such that, it represents the characteristic impedance of a line n connected to node n. The propagation constant matrix γ_{nn} and the transformation matrix T_{nn} are also defined in the same way. Z_{nm} is the load impedance matrix connected at the termination of the line m which is connected to node n. For $n = m = 1$, the situation represents the line connecting the node 1 to the source end.

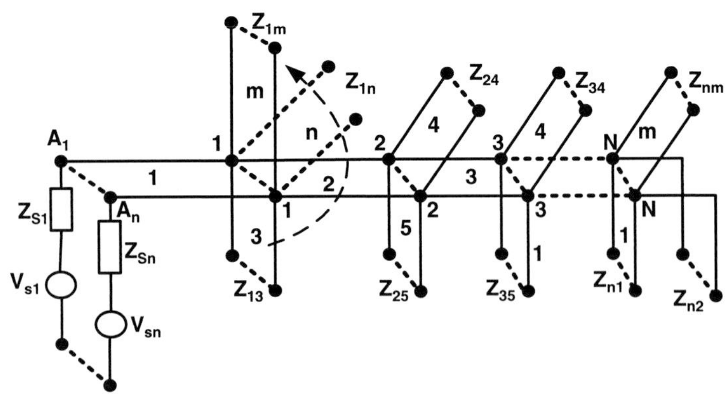

Figure 23: Power-line network with distributed branches.

$$H_{nm}(f) = Z_{C11}^{\ -1}\left(Z_{C11} + Z_S\right)a_{nm}A_{nm}\left(A_1^+\beta_{11} + A_1^-\right)^{-1} \tag{52a}$$

$$a_{nm}(f) = Z_{Cnm}\left(e^{-\gamma_{nm}L_{nm}}\beta_{nm} + e^{\gamma_{nm}L_{nm}}\right) \tag{52b}$$

$$A_{nm} = a_{mn}^{-1}\,a_n\,a_{(n-1)(n)}^{-1}\,a_{n-1}\,a_{(n-2)(n-1)}^{-1}\,a_{n-2}\ldots\ldots.a_{12}^{-1}a_1 \tag{52c}$$

$$\beta_{nm} = \left(T_{nm} + Z_{nm}^{\ -1}Z_{Cnm}T_{nm}e^{+\gamma_{nm}L_{nm}}\right)^{-1}\left(T_{nm} - Z_{nm}^{\ -1}Z_{Cnm}T_{nm}e^{-\gamma_{nm}L_{nm}}\right) \tag{52d}$$

$$\beta_{nn} = \left(C_{nn}^+ + P_nB_{nn}^+\right)^{-1}\left(C_{nn}^- - P_nB_{nn}^-\right) \tag{52e}$$

$$P_n = e_{nm(1)}a_{nm(1)}^{-1} + e_{nm(2)}a_{nm(2)}^{-1} + \ldots\ldots.. + e_{nm(M_T)}a_{nm(M_T)}^{-1} \tag{52f}$$

$$a_{nm} = B_{mn}^+\beta_n + B_{mn}^- \tag{52g}$$

$$e_{nm} = C_{mn}^- - C_{mn}^+\beta_n \tag{52h}$$

$$\beta_n = \begin{cases} \beta_{(n+1)(n+1)} & \text{to node } n+1 \\ \beta_{nm} & \text{to load } m \end{cases} \tag{52i}$$

$$a_{mn} = B_{nm}^+\beta_{nm} + B_{nm}^- \tag{52j}$$

$$a_n = B_{nn}^+\beta_{nn} + B_{nn}^- \tag{52k}$$

$$B_{mn}^+ = Z_{Cmn}T_{mn}e^{-\gamma_{mn}L_n} \tag{52l}$$

$$B_{mn}^- = Z_{Cmn}T_{mn}e^{\gamma_{mn}L_n} \tag{52m}$$

$$B_{nm}^+ = Z_{Cnm}T_{nm}e^{-\gamma_{nm}L_n} \tag{52n}$$

$$B_{nm}^- = Z_{Cnm}T_{nm}e^{\gamma_{nm}L_n} \tag{52o}$$

$$C_{mn}^+ = T_{mn}e^{-\gamma_{mn}L_n} \tag{52p}$$

$$C_{mn}^- = T_{mn}e^{\gamma_{mn}L_n} \tag{52q}$$

$$B_{nn}^+ = Z_{Cnn}T_{nn}e^{-\gamma_{nn}L_n} \tag{52r}$$

$$B_{nn}^{-} = Z_{Cnn} T_{nn} e^{\gamma_{nn} L_n} \tag{52s}$$

$$C_{nn}^{+} = T_{nn} e^{-\gamma_{nn} L_n} \tag{52t}$$

$$C_{nn}^{-} = T_{nn} e^{\gamma_{nn} L_n} \tag{52u}$$

In (52a), $A_1^{+} = (Z_{C11} + Z_S)T_{11}'$ and $A_1^{-} = (Z_{C11} - Z_S)T_{11}'$. The voltage response at any load impedance on line m connected to node n is given by (53).

$$V_{nm}(f) = H_{nm}(f)(Z_{C11} + Z_S)^{-1} Z_{C11} V_S(f) \tag{53}$$

8.4 Example validation of a generalized TL channel model for multi-conductor case using finite difference time domain (FDTD) method

The configuration with two coupled TL system as shown in Fig. 24 is considered. The line length of segments 1 and 2 are of 500 m long. The source and receiving end impedances were $Z_S = \begin{bmatrix} Z_{S1} & 0 \\ 0 & Z_{S2} \end{bmatrix} = \begin{bmatrix} 498 & 0 \\ 0 & 498 \end{bmatrix}$ and $Z_2 = \begin{bmatrix} Z_{21} & 0 \\ 0 & Z_{22} \end{bmatrix} = \begin{bmatrix} 50 & 0 \\ 0 & 50 \end{bmatrix}$.

The per unit length parameters, i.e. inductance was taken as $L_e = \begin{bmatrix} 1.686 & 0.599 \\ 0.599 & 1.686 \end{bmatrix} \mu\text{H/m}$ and capacitance $C_e = \begin{bmatrix} 7.844 & -2.874 \\ -2.874 & 7.844 \end{bmatrix} \text{pF/m}$ for line 1. The per unit length parameters, i.e. inductance was taken as $L_e = \begin{bmatrix} 0.7485 & 0.507 \\ 0.507 & 0.7485 \end{bmatrix} \mu\text{H/m}$ and capacitance $C_e = \begin{bmatrix} 34.432 & -18.716 \\ -18.716 & 34.432 \end{bmatrix} \text{pF/m}$ for line 2.

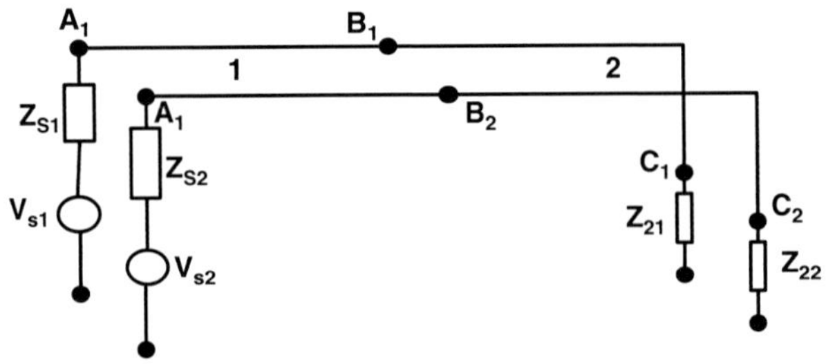

Figure 24: Coupled power-line network 1 terminated in another power-line network 2, both networks having different per unit length parameters.

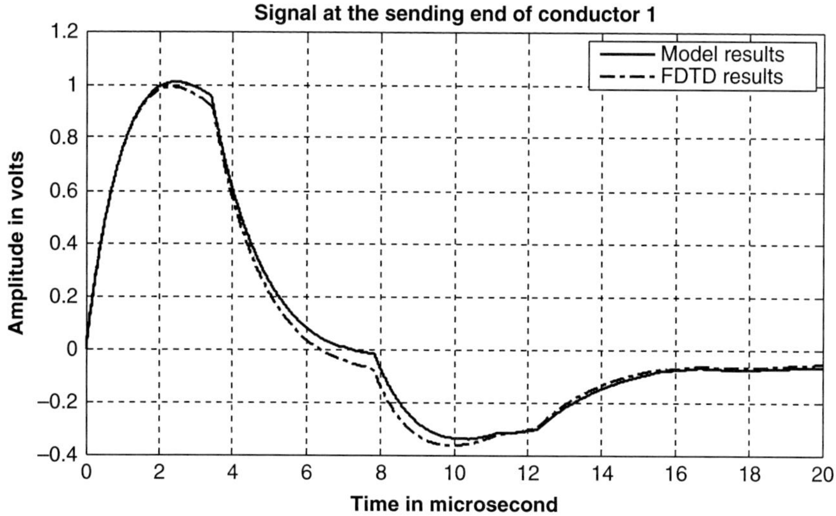

Figure 25: Responses at the sending end terminal of A1 based on generalized TL theory model and FDTD method.

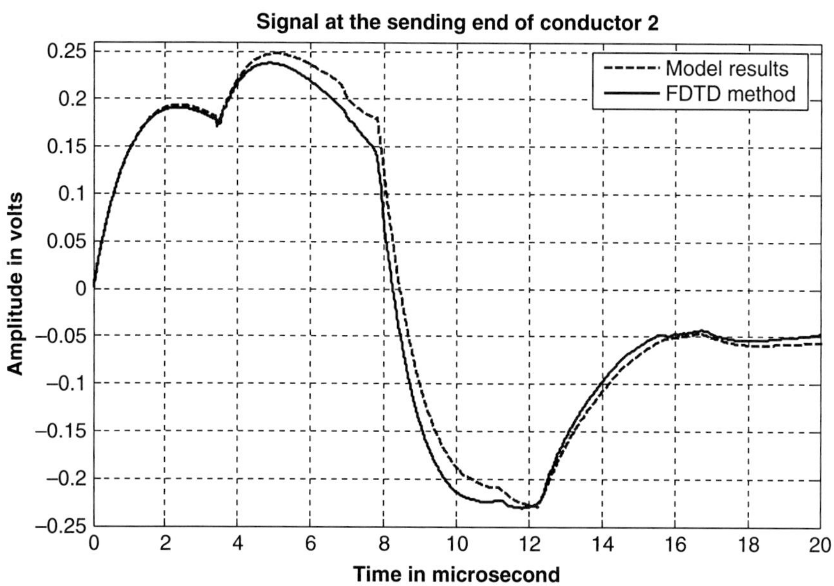

Figure 26: Responses at the sending end terminal of A2 based on generalized TL theory model and FDTD method.

The configuration was excited by a double exponential pulse, $V_{S1}(t) = 8.6(e^{-3\times10^5 t} - e^{-5.8\times10^5 t})$ which has amplitude of 2V on one of the conductor with V_{S2} (t) = 0. The time domain response was obtained by inverse Fourier transform of (53). The voltage at node A1 corresponding to the excited conductor is

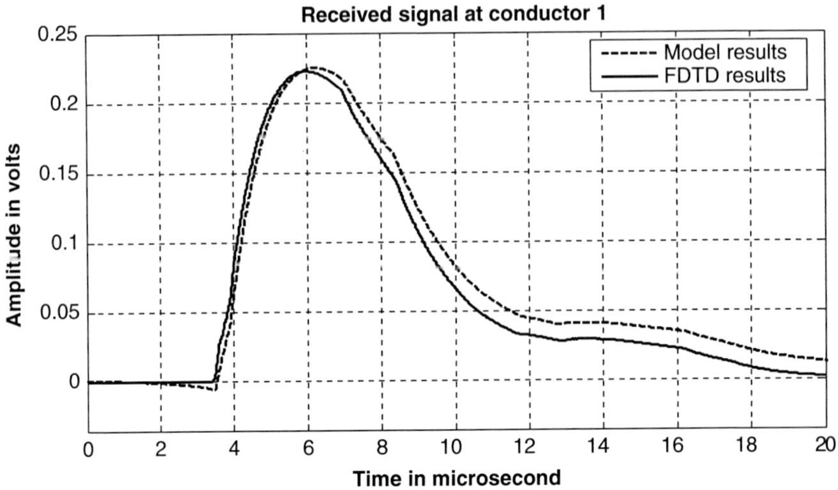

Figure 27: Responses at the receiving end terminal of C1 based on generalized TL theory model and FDTD method.

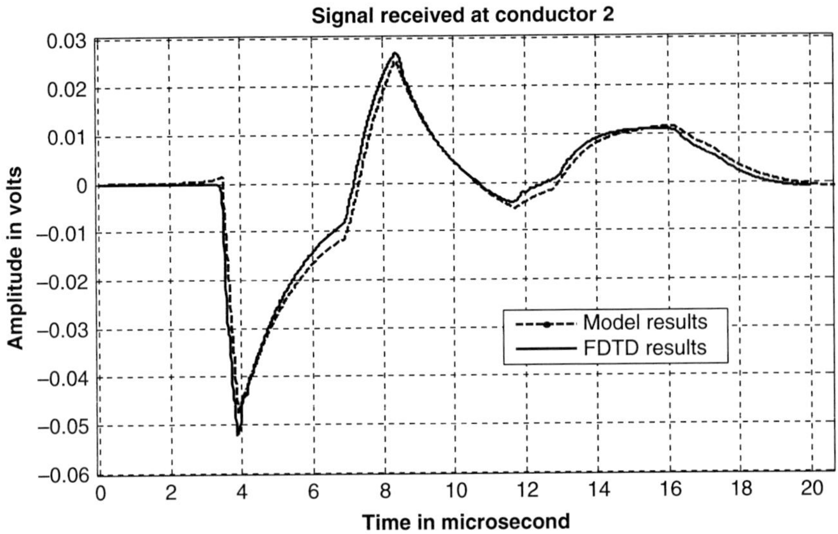

Figure 28: Responses at the receiving end terminal of C2 based on generalized TL theory model and FDTD method.

shown in Fig. 25, for the other conductor, i.e. A2 is shown in Fig. 26 using the proposed model and the FDTD method. The voltage at node C1 corresponding to the excited conductor is shown in Fig. 27, for the other conductor, i.e. C2 is shown in Fig. 28 using the proposed model and the FDTD method [14].

References

[1] Ezio, B. & Torino, P., Coding and modulation for a horrible channel. *IEEE Communications Magazine*, pp. 92–98, May 2003.

[2] Dommel, H.W., *Electromagnetic Transients Program (EMTP Theory Book)*, Bonneville Power Administration, 1986.

[3] Philipps, H., Modelling of powerline communication channels. *Proc. 3rd Intl Symp. Power-Line Communications and Its Applications*, Lancaster, UK, pp. 14–21, 1999.

[4] Hensen, C. & Schulz, W., Time dependence of the channel characteristics of low voltage power-lines and its effects on hardware implementation. *AEU Intl J. Electron. Commun.*, **54(1)**, pp. 23–32, February 2000.

[5] Zimmermann, M. & Dostert, K., A multipath model for the powerline channel. *IEEE Trans. Commun.*, **50(4)**, pp. 553–559, April 2002.

[6] Bewley, L.V., *Traveling Waves on Transmission Systems,* John Wiley & Sons, Inc., New York, 1933.

[7] Anatory, J., Kissaka, M.M. & Mvungi, N.H., Channel model for broadband powerline communication. *IEEE Transactions on Power Delivery*, **22(1)**, pp. 135–141, January 2007.

[8] Anatory, J., Theethayi, N., Thottappillil, R., Kissaka, M.M. & Mvungi, N.H., The effects of load impedance, line length and branches in the BPLC-transmission-line analysis for indoor voltage channel. *IEEE Transactions on Power Delivery*, **22(4)**, pp. 2150–2155, October 2007.

[9] Anatory, J., Theethayi, N., Thottappillil, R., Kissaka, M.M. & Mvungi, N.H., The effects of load impedance, line length and branches in the BPLC- transmission-line analysis for medium voltage channel. *IEEE Transactions on Power Delivery*, **22(4)**, pp. 2156–2162, October 2007.

[10] Anatory, J. & Theethayi, N., On the efficacy of using ground return in the broadband power line communications – a transmission line analysis. *IEEE Transactions on Power Delivery*, **23(1)**, pp. 132–139, January 2008.

[11] Anatory, J., Theethayi, N., Thottappillil, R., Kissaka, M.M. & Mvungi, N.H., Broadband power line communications: the channel capacity analysis. *IEEE Transactions on Power Delivery*, **23(1)**, pp. 164–170, January 2008.

[12] Anatory, J., Theethayi, N. & Thottappillil, R., Power-line communication channel model for interconnected networks – part I: two conductor system. *IEEE Transactions on Power Delivery*, **24(1)**, January 2009.

[13] Anatory, J., Theethayi, N. & Thottappillil, R., Power-line communication channel model for interconnected networks – part II: multi-conductor system. *IEEE Transactions on Power Delivery*, **24(1)**, January 2009.

[14] Paul, C.R., *Analysis of Multiconductor Transmission Lines,* John Wiley & Sons Inc., 1994.

[15] Anatory, J., Theethayi, N. & Mvungi, N.H., Power line channel models: comparisons between different modeling adopted in BPLC systems. WSPLC 2009 *Proceedings Third Workshop on Power Line Communications*, 1–2 October, Udine, Italy, pp. 35–38, 2009.

CHAPTER 4

The effects of line length, load impedance, number of branches in the BPLC

1 Introduction

In two-conductor transmission lines, there are cases whereby one conductor is carrying a signal in positive direction and other carrying signal in negative direction as discussed in the previous chapter. The conductor that is carrying current in negative direction is referred to as return conductor. There are cases whereby return conductor is either an adjacent transmission conductor or finitely conducting ground. In this chapter a case whereby adjacent conductor is used as return conductor is investigated and comparisons are made between the adjacent conductor return and the ground return. In general the model that will be used throughout from this chapter and the chapters to follow is the transmission line (TL) theory model as discussed in the previous chapter. All channel topologies like medium voltage (MV) channel, low voltage (LV) channel and indoor voltage (IV) channel with adjacent conductor as a return is considered. In addition some cases for a conductor with ground return for LV and MV channel are made for the sake of sensitivity analysis.

It has been observed that, there are a number of challenges associated with data transfer through power-line network infrastructure. Existing power-line topology (geometry and transmission voltage levels) varies from region to region and country to country. In countries like Tanzania, it has been observed that the MV, LV and IV systems exhibit a potential scope to extend the broadband services to end users. For MV systems typical direct line lengths between transmitting and receiving ends are about 4 km, with around 20 branches distributed along the line (interconnected or branched network configurations), with the connecting branch lengths ranging from 30 to 100 m. Moreover, the terminal loads experienced by such line configuration may not be always having impedance equal to the line characteristic impedance or resistive loads. For LV systems direct line lengths between transmitting and receiving ends are about 1.2 km (underground cables and overhead transmission lines). For IV systems the maximum direct line lengths between transmitting and receiving ends are about 20 m. It is thus appropriate that with such different channel topologies a number of case studies are to be carried out so as to provide guidelines for a possible future optimal planning and design of communication systems.

Different studies regarding the effect of load impedances, branches, etc. have been reported by Mathias *et al.* [1], Pavlidou *et al.* [2], Zimmermann *et al.* [3], etc. Zimmermann *et al.* [3] and Mathias *et al.* [1] pointed out that maximum distance for a lossless data transmission through power line is about 300 m. Pavlidou *et al.* [2] concludes that "studies are still necessary to better understand and improve the performance of power lines for higher bit rate transmission". Researchers have investigated the variation in time/frequency responses due to the influence of load impedance, line length and branches without mentioning exactly/clearly, what was the contribution of each parameter to the stochastic behavior of channel responses. For example, the answers to the following questions should draw some conclusion for the performance improvement of MV, LV and IV communication channels:

- How does number of branches contribute to the total signal response?
- How does line lengths (d_L) from transmitter to the receiver and branched line length (X) influence the signal response?
- How do the terminal load (infinite and low) impedances (Z_{inf} and Z_{Low}) influence the signal response?
- How finitely conducting ground return influence signal response?

To the authors' knowledge in the literature, no complete systematic study was made to address all the above questions together, perhaps due to the complexities and uncertainties involved in the adopted channel modeling. Consequently, through this chapter an attempt is made to answer the above questions. In the analyses to be presented the frequency response (or the transfer function as applicable) is calculated based on TL theory model developed by Anatory *et al.* [4].

2 Medium voltage channel

2.1 Effects of line lengths

2.1.1 Effects of length from transmitter to the receiver

A typical medium voltage line of Tanzania power network is considered. All the lines have the following per unit length parameters with (L_e = 1.9648 μH/m, C_e = 5.6627 pF/m) [5]. The configuration under study is given in Fig. 1 with $Z_L = Z_S$ = 589 Ω. The line length AC was varied as 4 km, 2 km, 1 km and 500 m, with length of BD constant at 40 m. In the study here B is always the mid point of AC. Point D was terminated in 50 Ω.

Figure 2a–d shows frequency responses of transfer functions relating the load voltages at C and the sending end as given in the previous chapter for 4 km, 2 km, 1 km and 500 m, respectively. From Fig. 2, the peak values of signal response were not attenuating significantly with either frequency or line length. The position of notches in the signal response of medium voltage channel also does not depend on

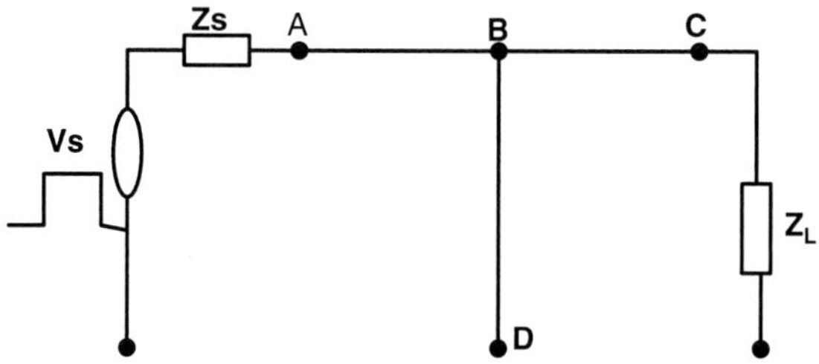

Figure 1: Power-line network with a branch.

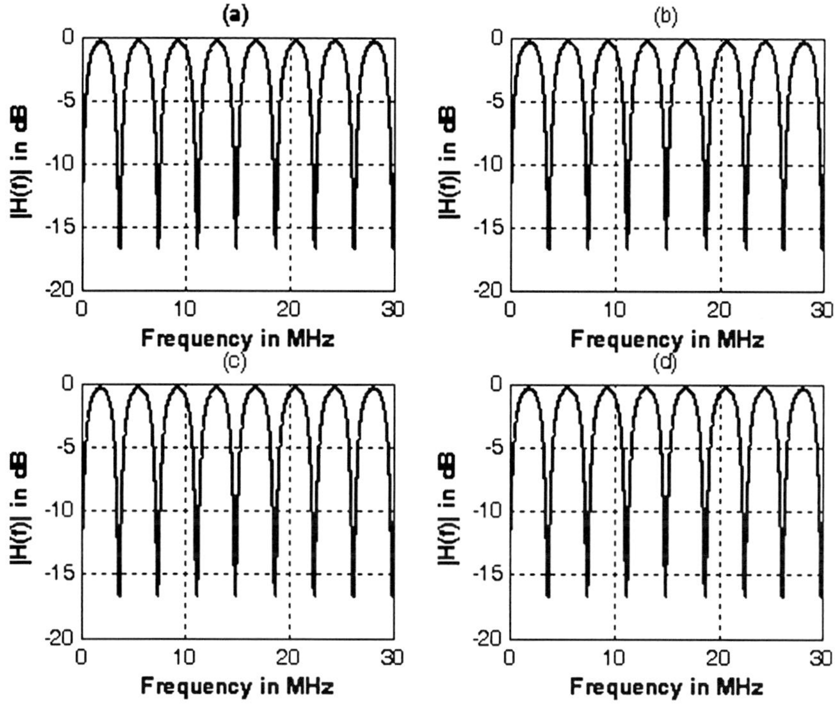

Figure 2: Simulation results for medium voltage power-line link with one branch (a) 4 km (b) 2 km (c) 1 km (d) 500 m.

length from transmitter to receiver. Figure 3a–d shows the corresponding phase responses. It is observed that as the line length increases there are rapid changes in the phase response. This could perhaps limit the available transmission bandwidth of the medium voltage channel.

Figure 4a–d is the received signal for a transmitted rectangular pulse with amplitude 2 V (pulse width of 1 μs and shifted by 0.5 μs) for different line lengths. From the received pulse for all cases the amplitude is fluctuating between

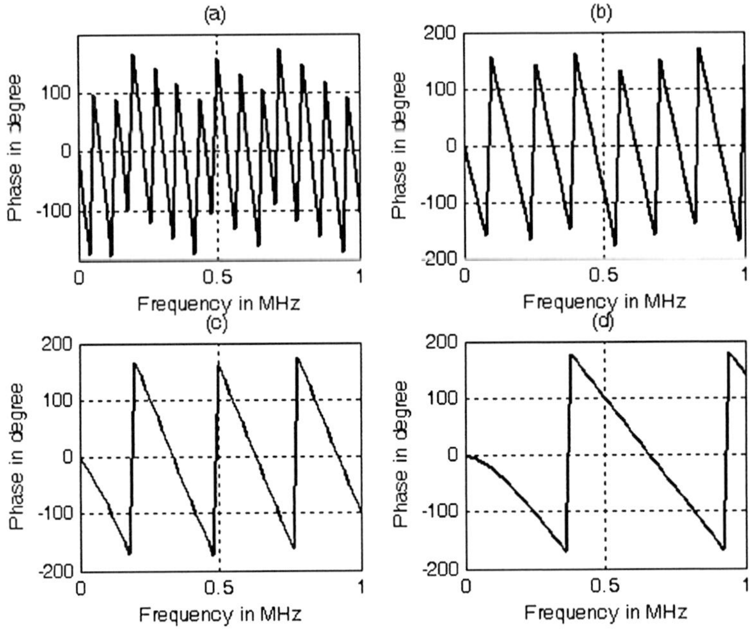

Figure 3: Phase response for medium voltage power-line link with one branch (a) 4 km (b) 2 km (c) 1 km (d) 500 m.

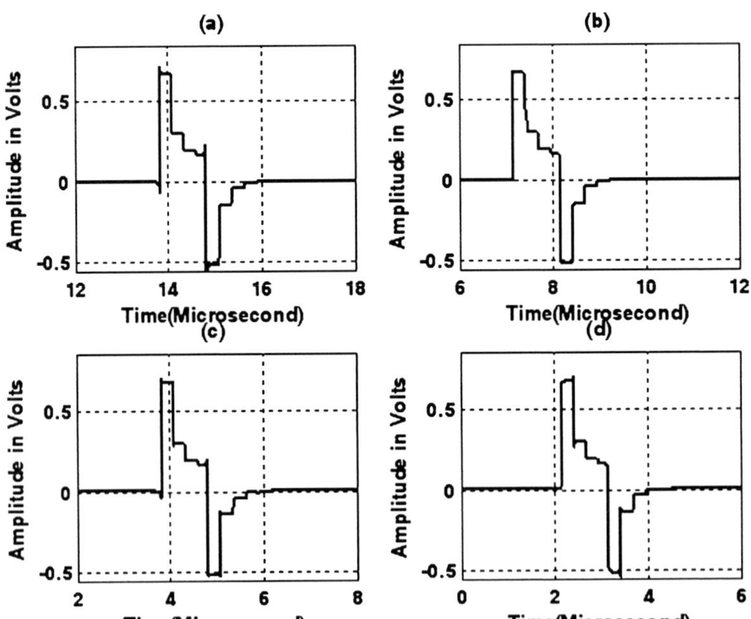

Figure 4: Received pulses at point C in Fig 1; for medium voltage power-line link with one branch (a) 4 km (b) 2 km (c) 1 km (d) 500 m.

0.7 and –0.5 V with distorted shapes. This shows that in medium voltage channel, signals encounter both attenuation and distortions which do not depend on transmission line length, mainly due to length of interconnected branches and its associated terminal loads.

2.1.2 Effects of branch length

The configuration as given in Fig. 1 was used; i.e. the length of a line from point A to C was kept constant at 4 km. The branch length was varied as BD = 10, 20, 40 and 80 m with B always at the mid of line AC. Point D was terminated in 50 Ω as in the previous case and the same exercise was repeated as before, i.e. calculating the transfer characteristics with respect to the load at C. Figure 5a–d shows the corresponding frequency responses for various branch line lengths. It is observed that, in all cases the peaks of frequency responses were not either attenuating with frequencies or branch length similar to the earlier case. Whereas, the position of the peaks and notches is case dependent unlike the previous case. The generalized expression for frequency position (f_i in MHz) of the ith peak in terms of branched line length (X in m) is approximately given by (1). Similarly, the positions of the notches are given by (2). As the length of branched line increases the number of notches increases. The phase response for the case under study had similar behavior as in Fig. 3a. It appears that the

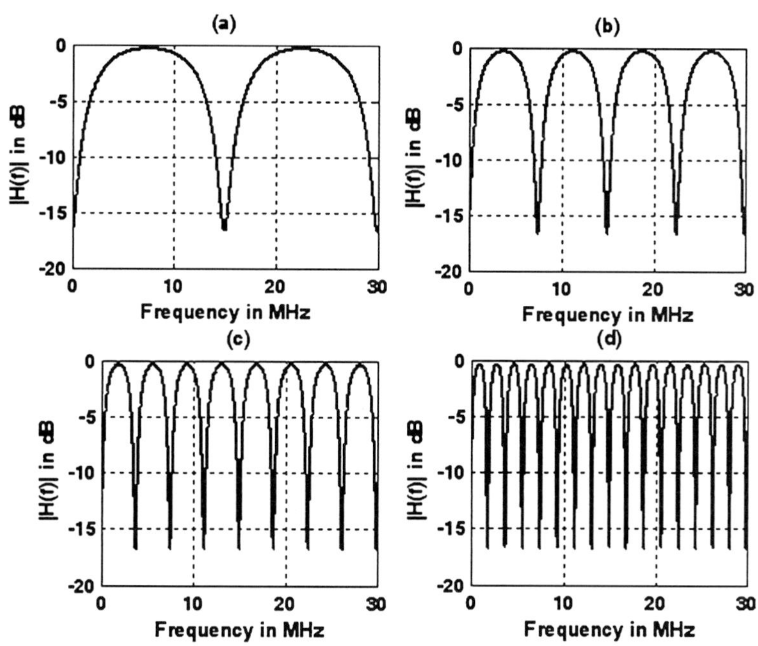

Figure 5: Simulation results for medium voltage channel of 4 km with one branch of length (a) 10 (b) 20 (c) 40 (d) 80 m.

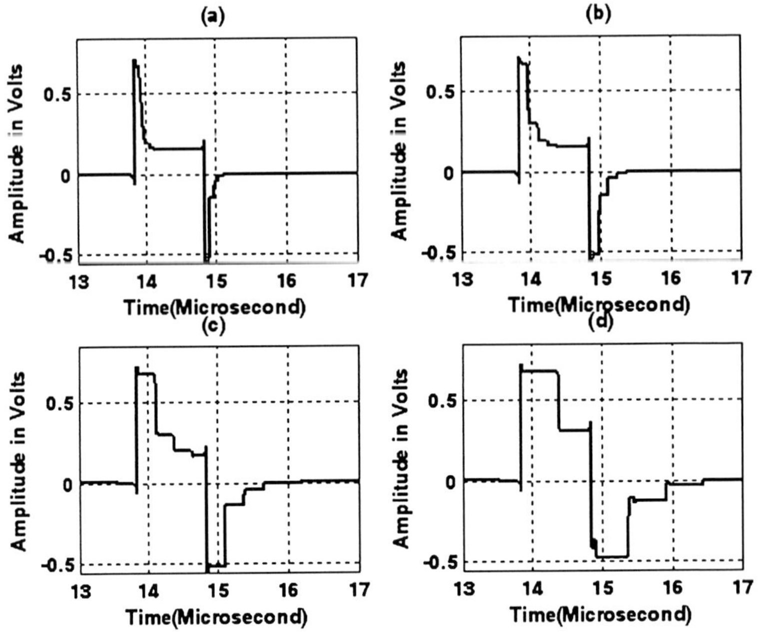

Figure 6: Received pulses at point C in Fig 1; for medium voltage power-line link with one branch of length (a) 10 (b) 20 (c) 40 (d) 80 m.

length of branched transmission line still does not affect the phase response of MV channel.

$$f_i = \frac{71.7}{X}(1+2i), \quad \forall i = 0,1,2... \tag{1}$$

$$f_i = \frac{145.5}{X}\cdot i, \quad \forall i = 0,1,2... \tag{2}$$

Figure 6a–d shows the received time domain signal for a rectangular pulse injected at the sending end similar to the previous cases for various branch lengths. From the results it can be observed that the received signal has similar peak-to-peak characteristics compared to earlier case. As the branched length becomes shorter the received signal becomes more distorted.

2.2 Effects of number of branches

Consider an MV channel with distributed branches as shown in Fig. 7. The number of branches was varied in the link between points A and J. The distance between points A and J was 4 km, while all branches were 40 m long. The number of branches was varied as 2, 4 and 8. Note that for each case the branches were equally distributed between points A and J. The terminations of all the branches

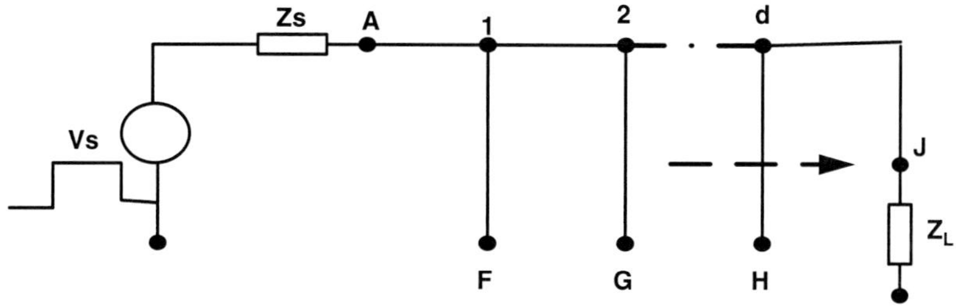

Figure 7: Power-line medium voltage network with distributed branches.

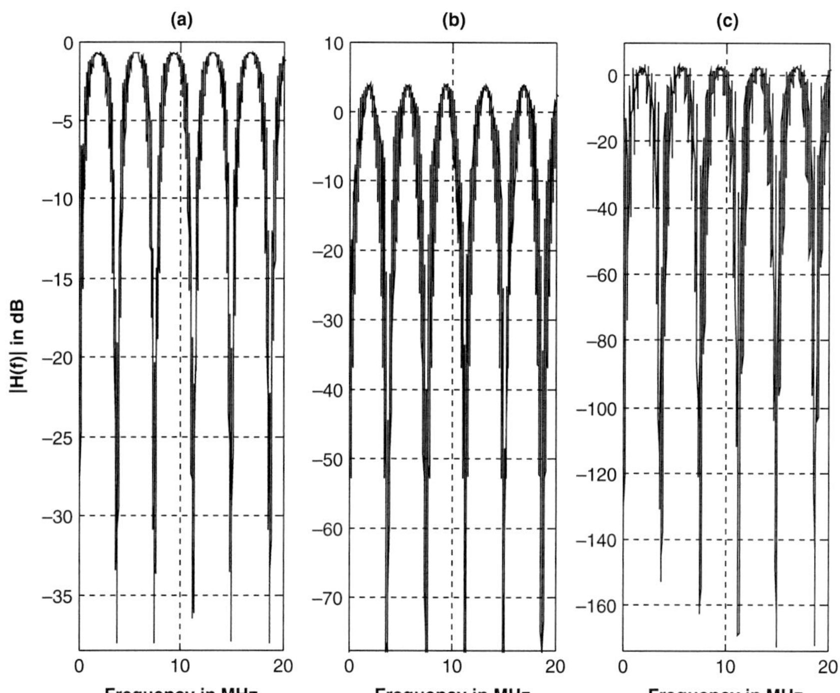

Figure 8: Simulation results for medium voltage channel with distributed branches (a) 2 branches, (b) 4 branches (c) 8 branches.

were 50 Ω. Figure 8a–c shows the corresponding frequency responses for different number of branches. It is observed that the positions of notches are not changed. But as the number of branches increases the attenuations of notched point tend to increase.

The phase responses were comparable to previous case as shown in Fig. 3a. Figure 9a–c shows the received time domain signals for the same injected source at the transmitting end as in the previous case. For the case of two distributed branches the signal peak-to-peak voltage was between 0.5 and –0.5 V; similarly for 4 branches the signal peak-to-peak voltage was between 0.4 and –0.4 V. For 8

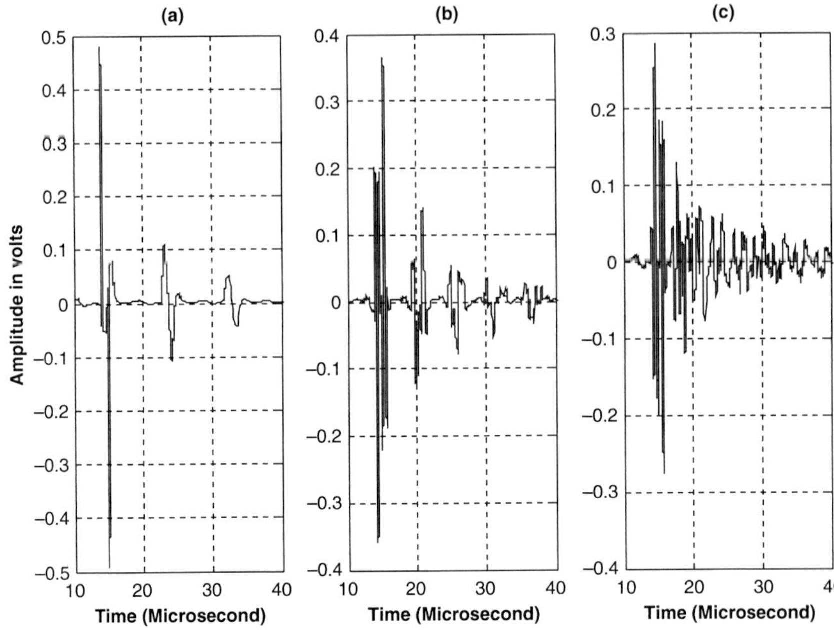

Figure 9: Received signals at point J in Fig 7; for medium voltage channel with distributed branches: (a) 2 branches, (b) 4 branches (c) 8 branches.

branches the signal peak-to-peak voltage was 0.3 V. This indicates that in medium voltage channel as the number of branches increases it creates both attenuations and severe signal distortions.

2.3 Effects of load impedance

This study is emphasized here because it is common that the loads at the termination of branched lines are not always equal to line characteristic impedance or resistive. Instead the channel is terminated in arbitrary load, like, low or high impedance i.e., resistive (R type) compared to line characteristic impedance and practical load impedance i.e., resistive-inductive (RL type) representing transformers, machines, etc. For discussions, the configuration as in Fig. 1 is considered. The length of line AC was kept constant and equal to 4 km, while branch BE of length 40 m was connected to the middle of line AC. The termination of point D was varied according to the given load impedance under investigation. Note that Z_S and Z_L are the characteristic impedances of the line AC.

2.3.1 Resistive loads

The following load impedances with values of 4 Ω, 40 Ω, 400 Ω, 589 Ω, 4 kΩ and 40 kΩ terminated at D were considered. Note 589 Ω is the characteristic impedance of the line BC. Figure 10a–d shows the frequency responses for medium

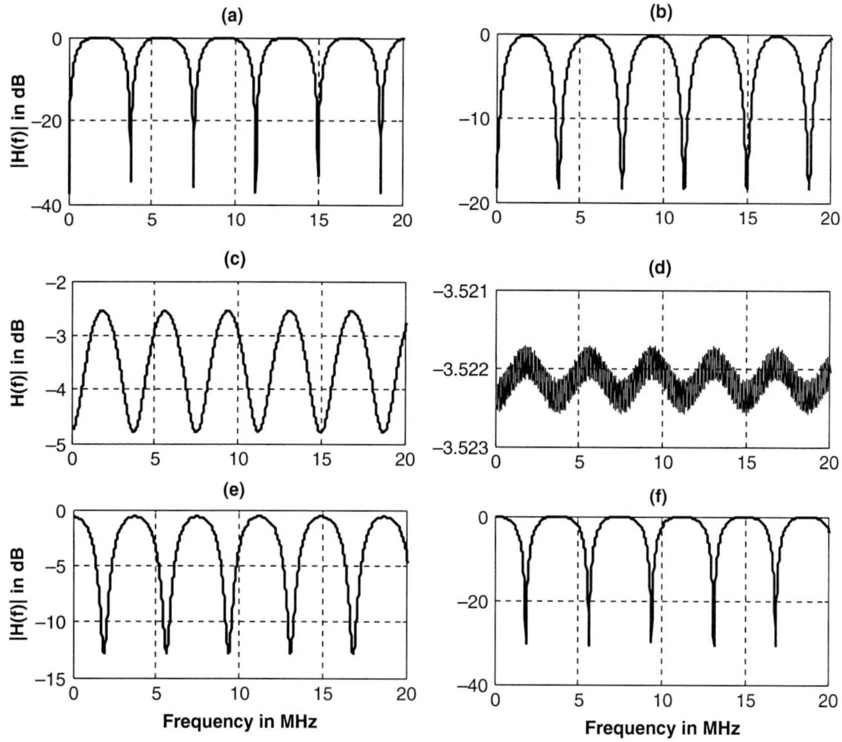

Figure 10: Results for a medium voltage channel with a branch terminated in low impedances: (a) 4 Ω (b) 40 Ω (c) 400 Ω (d) 589 Ω (e) 4 kΩ (f) 40 kΩ.

voltage channel for various termination impedances at D. For the load impedances less than channel characteristic impedance the position of notches is unchanged with no attenuation (see Fig. 10a and b). It is interesting to observe that when the load impedance is low, the peaks are at 0 dB and the notches are at –40 dB. As the load impedance increases the peaks increase and the notches decrease. As the load is equal to the characteristic impedance, the peaks and notches disappear. When the load impedance increases beyond the characteristic impedance, the peaks and notches behave in the same way as if it were approaching lower impedances, but with a shift in their frequency positions.

The generalized expression for the frequency positions of notches for the load impedance terminated in impedance less than line characteristic impedance is given by (3) while for termination impedance greater than line characteristic impedance is given by (4). Similarly, the position of peak frequency for load impedance less than line characteristic impedance is given by (3), while for load impedance greater than line characteristic impedance is given by (4). The phase responses in the frequency ranges 0–1 MHz has similar features as in Fig. 3a.

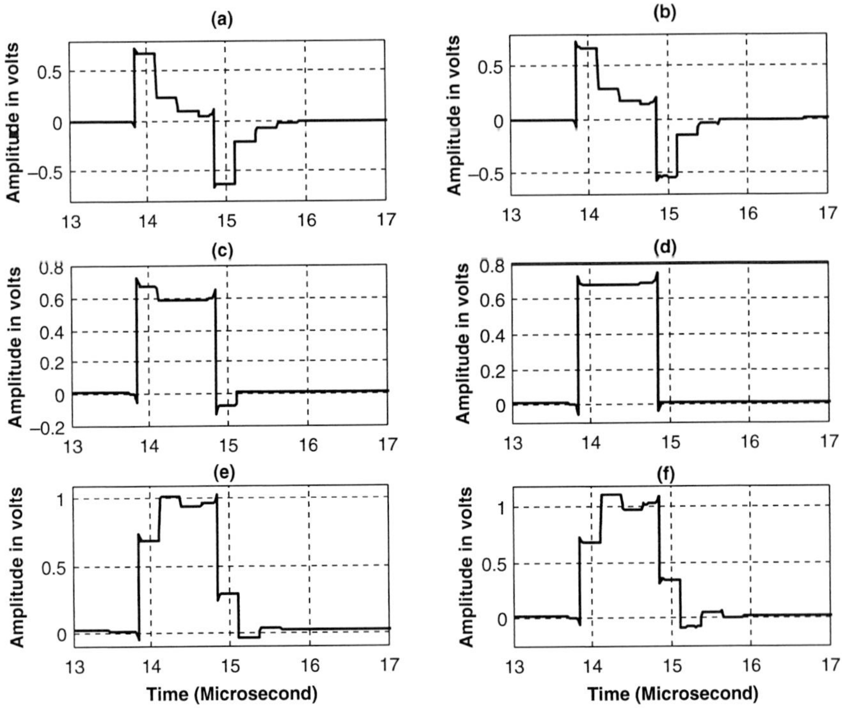

Figure 11: Received signals with a branch terminated in low impedances at point C in Fig 1; (a) 4 Ω (b) 40 Ω (c) 400 Ω (d) 589 Ω (e) 4k Ω (f) 40k Ω.

$$f_i = \frac{72.9}{X}(1+2i), \quad \forall i = 0,1,2... \tag{3}$$

$$f_i = \frac{144}{X} \cdot i, \quad \forall i = 0,1,2... \tag{4}$$

Figure 11a–f shows the received signals for a medium voltage channel with one branch and terminated in 4 Ω, 40 Ω, 400 Ω, 589 Ω, 4 kΩ and 40 kΩ, respectively, for an injected 2-V rectangular pulse with pulse width of 1 μs, shifted by 0.5 μs. It can be observed that as the load impedance increases from lower to higher values, both signal attenuation and distortions tends to reduce.

2.3.2 Inductive loads

Now consider an RL load termination at terminal D of Fig. 1, where the inductance was varied as 0.005, 0.05, 0.5, 5, 50 and 500 mH with constant resistance of 50 Ω. The frequency response is as shown in Fig. 12. It can be observed that the inductive load behavior shift from short circuit to open circuit behavior as shown in Fig. 12. The phase response, minor differences were observed compared to Fig. 3a. Figure 13a–d shows the received signals for a 4-km long medium voltage channel with one branch terminated in 0.005, 0.05, 0.5, 5, 50 and 500 mH,

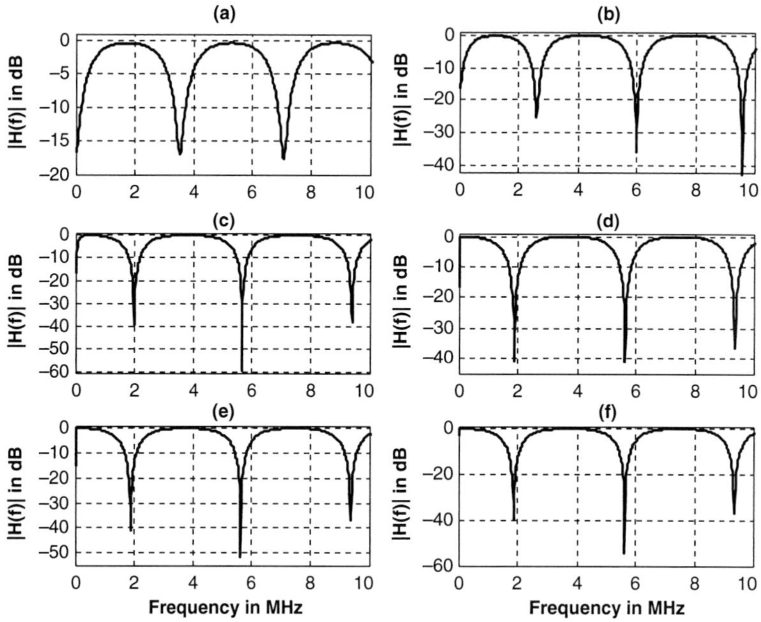

Figure 12: Frequency response for medium voltage channel with a branch terminated in inductive load with (a) 0.005 mH (b) 0.05 mH (c) 0.5 mH (d) 5 mH (e) 50 mH (d) 500 mH.

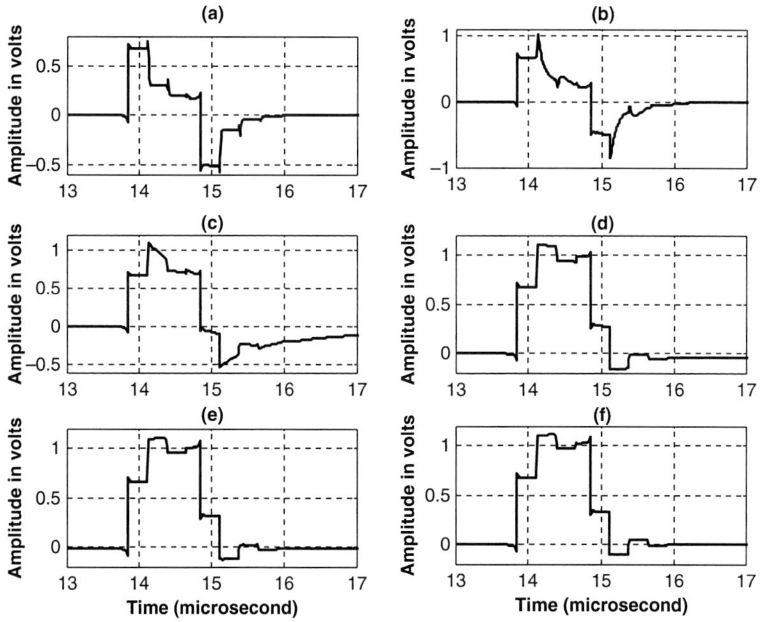

Figure 13: Received signals for medium voltage channel with a branch terminated in inductive load at point C in Fig 1 with : (a) 0.005 mH (b) 0.05 mH (c) 0.5 mH (d) 5 mH (e) 50 mH (d) 500 mH.

respectively, for the same voltage source as used in the previous case. It can be observed that as the inductance tends to be lower the signal distortion increases.

3 Low voltage channel

3.1 Effects of line length

3.1.1 Length from transmitter to the receiver

A typical low voltage line of Tanzanian power network with line parameters as L_e = 1.9589 μH/m, C_e = 5.6799 pF/m is considered [6]. The configuration is shown in Fig. 1, with $Z_L = Z_s$ = 587 Ω. The line length AC was varied as 1.2 km, 600 m, 300 m and 150 m, with point B always at mid point of AC. The branched line length (BD) was considered constant equal to 30 m and terminated in 50 Ω. Figure 14a–d shows the transfer functions relating the load and sending end voltages for various lengths of AC. From Fig. 14, the peaks and notches in frequency response do not vary with either frequency or line length. The positions of peaks and notches are independent of the line length from transmitter to receiver.

Figure 15a–d shows the corresponding time domain impulse responses for various lengths of AC. The first peak, in each figure, represents the direct path and the rest are reflected paths. It can be observed that the direct path attenuates as length increases. Figure 16a–d is the received signal for a sending end pulse having 2-V peak, width of 0.5 μs, rise time 1 ns, shifted by 0.5 μs. Looking at all signals, it is seen that the pulse suffered attenuation.

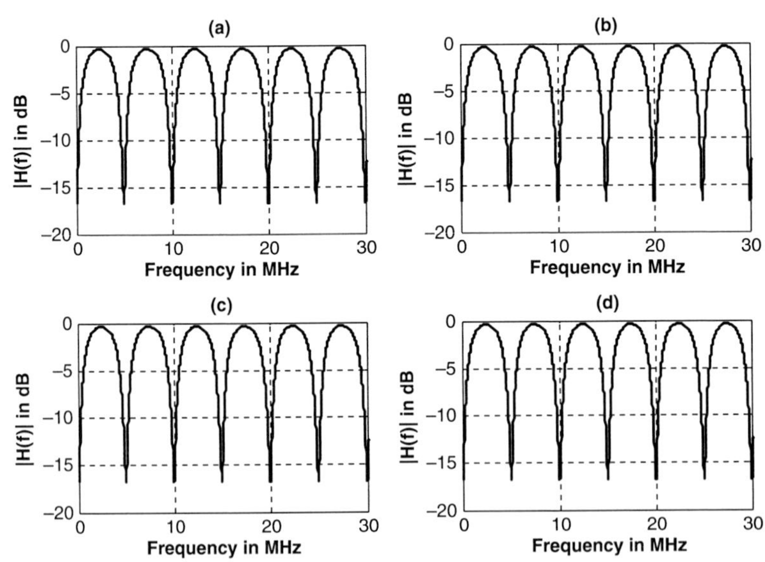

Figure 14: Simulation results for low voltage power-line link with one branch of (a) 1.2 km (b) 600 m (c) 300 m (d) 150 m.

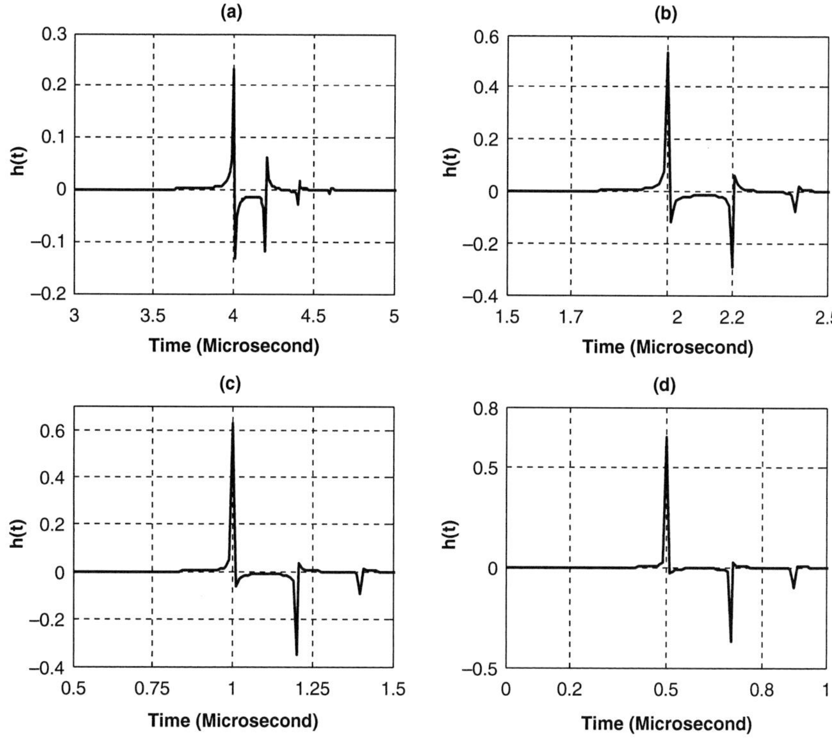

Figure 15: Impulse response for low voltage power-line link with one branch line length of (a) 1.2 km (b) 600 m (c) 300 m (d) 150 m.

3.1.2 Branched length

The configuration is the same as in Fig. 1, but the length of AC was kept constant at 1.2 km, while the length of BD was varied as 10, 15, 20 and 30 m. Point D was terminated in 50 Ω. The transfer functions for all cases relating the voltages at the load and launched voltages at point A are shown in Figure 17a–d. It can be observed that the positions of notches and peaks are case dependent, i.e. dependent on branched line length. As the branched line length increases, it results in more notched points. The generalized expression for frequency position (f_i in MHz) of the ith peak in terms of branched line length (X in m) is approximately given by (5). Similarly the positions of the notches are given by (6).

$$f_i = \frac{75}{X}(1+2i), \quad \forall i = 0,1,2... \tag{5}$$

$$f_i = \frac{150}{X} i \ \ \forall i = 0,1,2... \tag{6}$$

Figure 18a–d shows the corresponding time domain impulse responses for different branch lengths. Looking at the amplitude of direct path was constant at

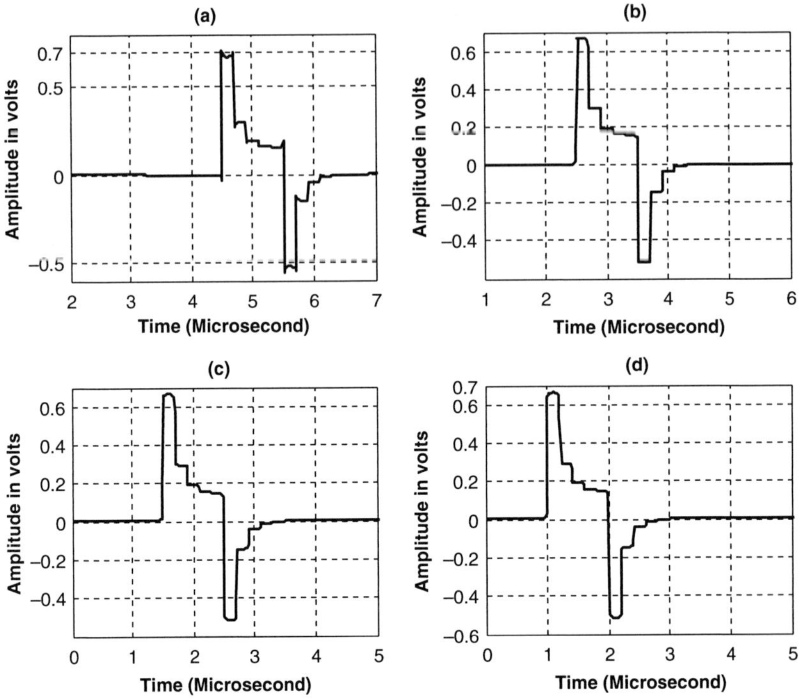

Figure 16: Received pulses at point C in Fig 1; for low voltage power-line link with one branch of (a) 1.2 km (b) 600 m (c) 300 m (d) 150 m.

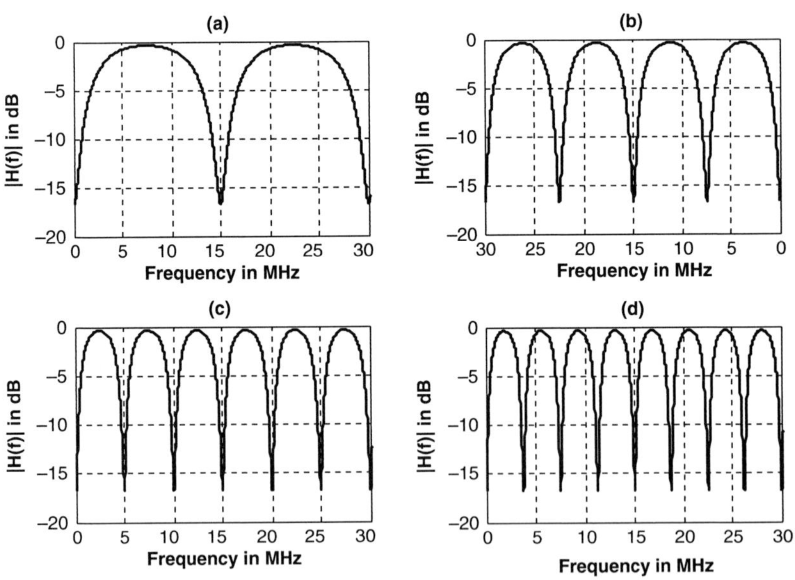

Figure 17: Simulation results for low voltage channel of 1.2 km with one branch of length (a) 10 m (b) 20 m (c) 30 m (d) 40 m.

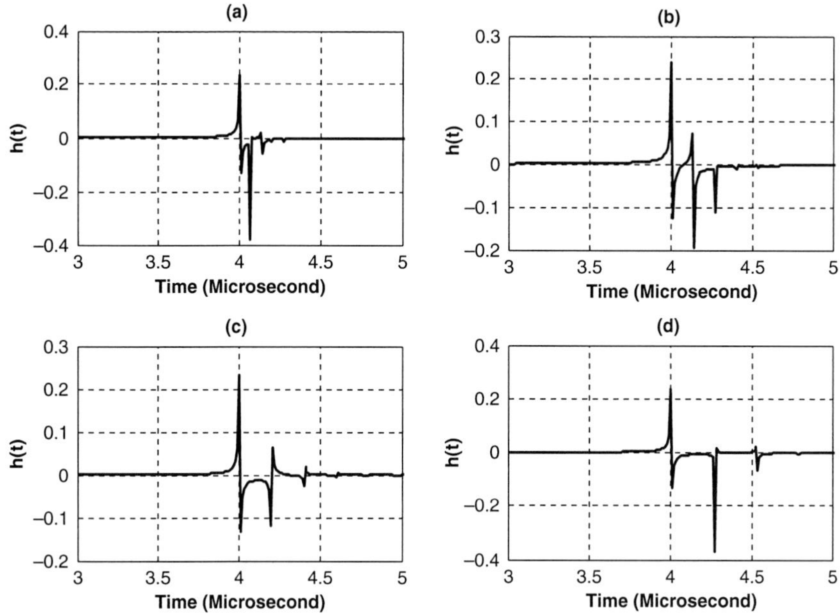

Figure 18: Impulse response simulation results for low voltage channel of 1.2 km with one branch of length (a) 10 m (b) 20 m (c) 30 m (d) 40 m.

about 0.25, while the second reflected path's amplitude was fluctuating depending upon branched line length.

3.2 Effects of number of branches

3.2.1 Multiple branches at single node

Consider the configuration as shown in Fig. 19, with Z_s and Z_l same as before. The length of AC is 1.2 km with all branches 20 m long concentrated at node B. All terminal loads were terminated in 50 Ω. The numbers of branches are varied as 2, 4, 8 and 16. Figure 20a–d shows the transfer functions for all cases relating the voltages at the load and launched voltages at point A. For 2 branches (Fig. 20a) the peaks are –2 dB, while notches are –22 dB. For 4 branches (Fig. 20b) the peaks' attenuations of frequency response vary as –4 dB, while notches are –28 dB. With 8 and 16 branches the notches are about –32 and –40 dB, respectively. For 8 and 16 branches (see Fig. 20c and d) the peak is between –4 and –5 dB, respectively.

Figure 21a–d shows the time domain impulse responses of the channel with multiple branches at the same node. It is observed that as the number of branches increases the direct path tend to be more attenuated and the reflected paths do not vary much (Figs 21 and 22). However, if the reflected paths gain larger magnitude severe channel distortions could occur. Figure 22a–d shows the received signal for 2V rectangular signal with pulse width 1 μs shifted by 5 μs for various number of branches. It can be observed that in low voltage channel, increasing the number of

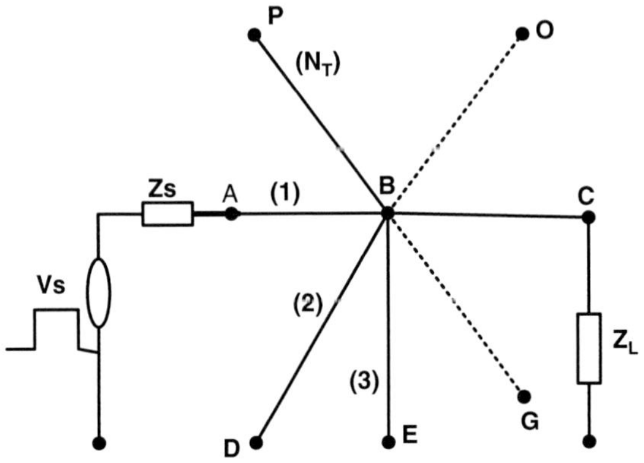

Figure 19: Power-line network with multiple branches at a single node.

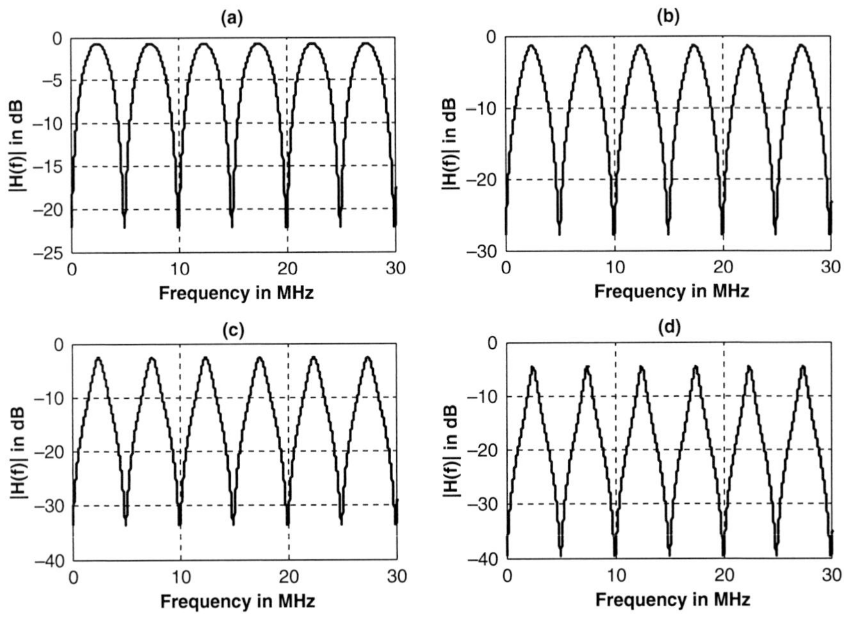

Figure 20: Simulation results for low voltage channel with multiple branches at single node (a) 2 branches (b) 4 branches (c) 8 branches (d) 16 branches.

branches at the same node causes the received signals to be more attenuated and distorted.

3.2.2 Distributed branches

Consider the low voltage channel with distributed branches as shown in Fig. 7. The number of branches was increased in the link between point A and J. The length

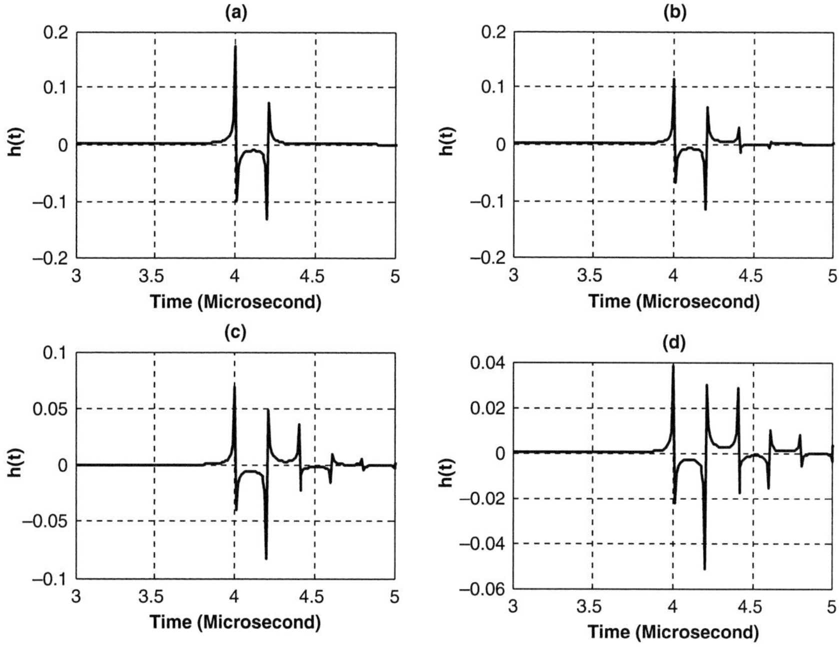

Figure 21: Impulse response for low voltage channel with multiple branches at single node (a) 2 branches (b) 4 branches (c) 8 branches (d) 16 branches.

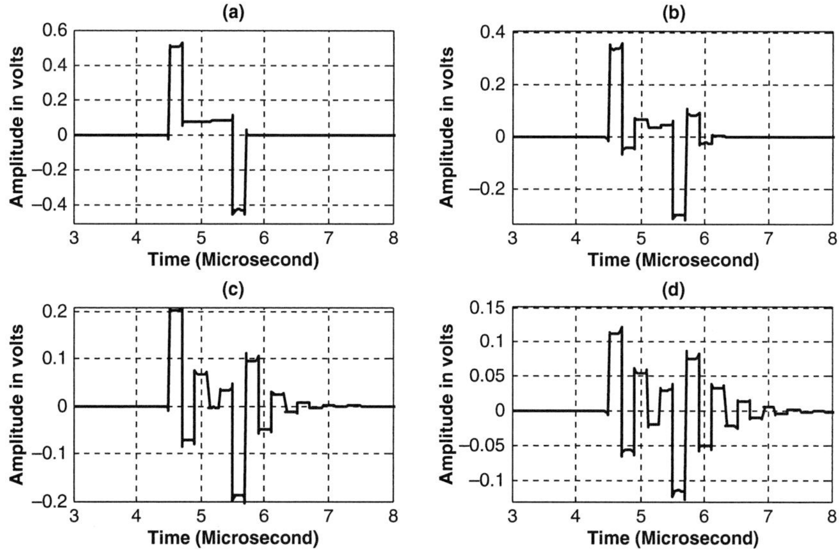

Figure 22: Received signal at point C in Fig 19 for low voltage channel with multiple branches at single node (a) 2 (b) 4 (c) 8 (d) 16 branches.

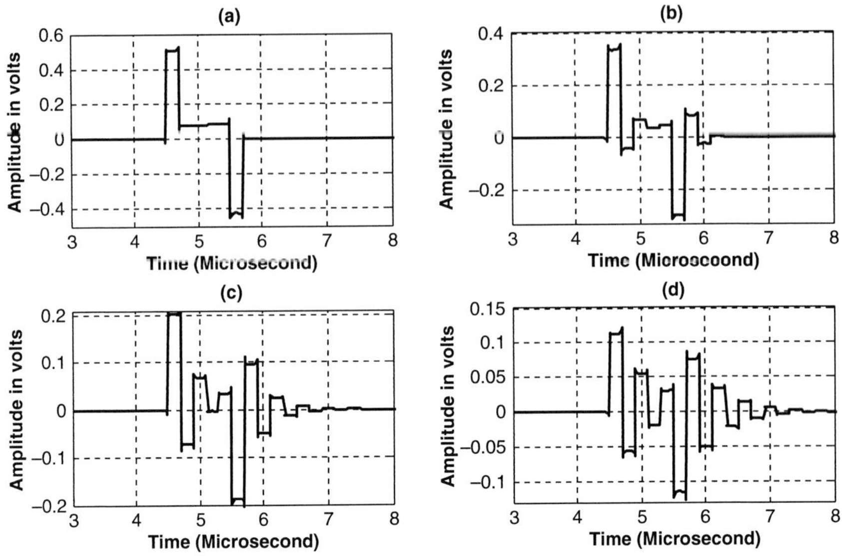

Figure 23: Simulation results for low voltage channel with distributed branches (a) 2 (b) 4 (c) 8 branches.

AJ was 1.2 km while all branches were 20 m. The number of branches was varied as 2, 4 and 8 such that in each case they were equally distributed. The terminations were considered as 50 Ω. Figure 23a–d shows the transfer functions for all cases relating the voltage at the load and launched voltage at point A for different number of branches. It can be seen that the positions of deep notches are not changed. As the number of branches increase the attenuations of notched point tends to increase.

The stochastic attenuations are observed as the number of branches increases which increases the possibilities of reducing the available bandwidth for low voltage channel. Figure 24a–d shows the corresponding time domain impulse responses. It can be observed that as the number of distributed branches increases, it leads to multiple reflections from all available paths, which will require advanced equalization techniques for signal recovery. Figure 25a–d shows the received signals for a low voltage channel of 1.2 km with different number of branches. It is evident that increase in the number of branches leads to multiple reflections causing additional attenuated peaks. Larger distortions are observed with increase in the number of branches.

3.3 Effects of load impedance

3.3.1 Low resistive load

The effects of load impedances were considered for the configuration shown in Fig. 1. The load impedances at point D were varied as 1 Ω, 100 Ω, 200 Ω and 400 Ω.

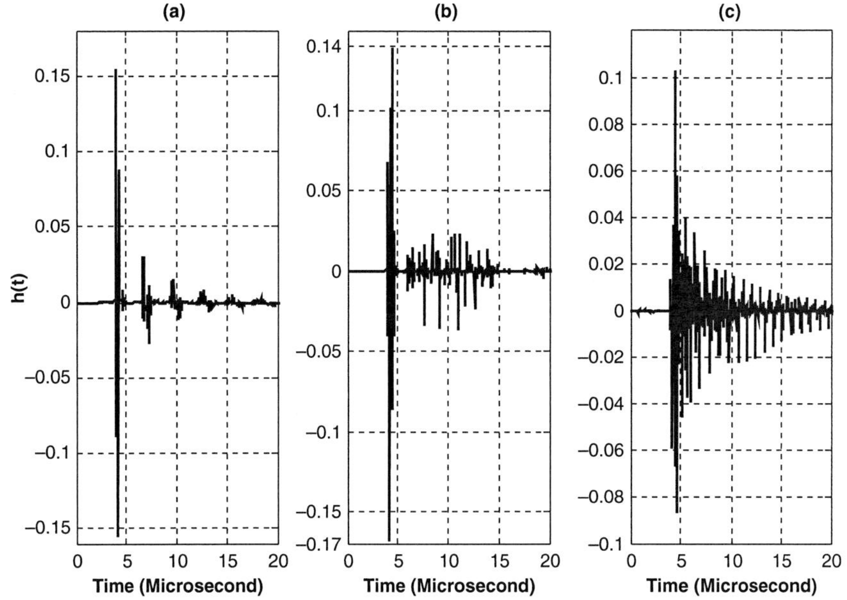

Figure 24: Impulse response for low voltage channel with distributed branches (a) 2 (b) 4 (c) 8 branches.

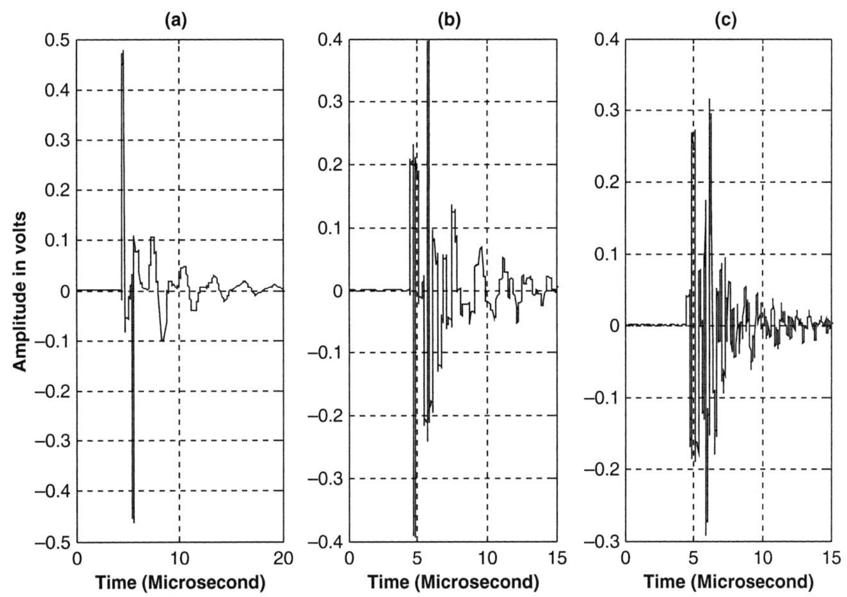

Figure 25: Received pulse at point J corresponds to Fig 7; for low voltage channel with distributed branches (a) 2 (b) 4 (c) 8 branches.

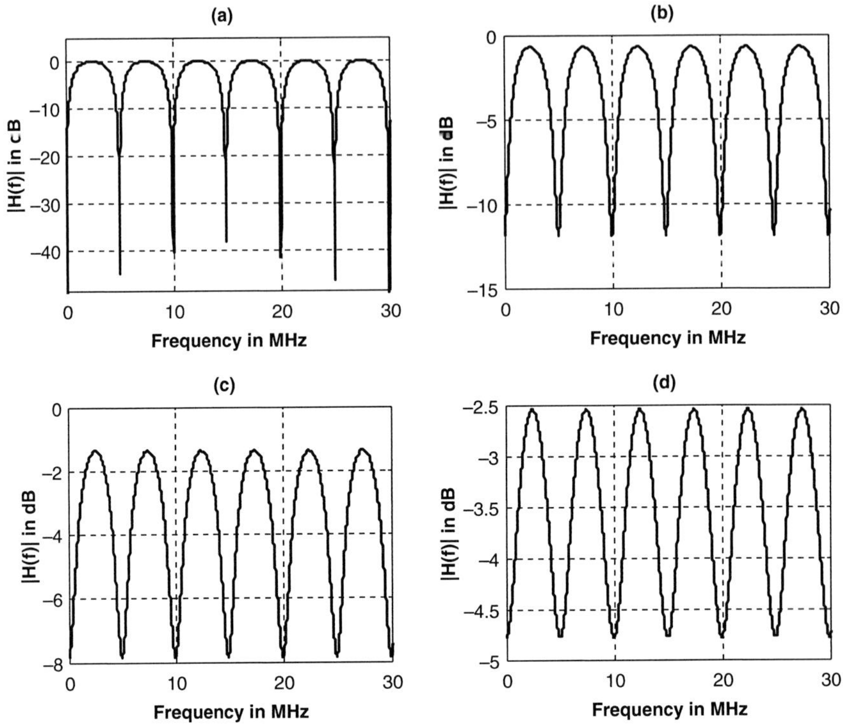

Figure 26: Simulation results for low voltage channel with a branch terminated in: (a) 1 Ω (b) 100 Ω (c) 200 Ω (d) 400 Ω.

The length of AC was 1.2 km and BD was 20 m. Figure 26a–d shows transfer functions for all cases relating the voltage at the load and launched voltage at point A for different load conditions. It can be seen that as the load impedance increases towards characteristic impedance, the peak of frequency responses tends to decrease more and the notches tend to improve. Figure 27a–d shows the time domain impulse responses of the channel. It can be observed that for all cases, the direct path have the same amplitude of about 0.25. As the load impedance increases towards the characteristic impedance the successive reflected path tends to attenuate more. Figure 28a–d shows the received signals for a low voltage channel with one branch and terminated in various load impedances, respectively, for a sending end signal of 2-V rectangular pulse, of width 1 μs, shifted by 0.5 μs. It can be observed that as the load impedance increases, both signal attenuation and distortions tend to decrease.

3.3.2 High resistive load

The configuration given in Fig. 1 is considered with high resistance load variation as 600 Ω, 10 kΩ, 100 kΩ and 10 MΩ. Figure 29a–d shows transfer functions for

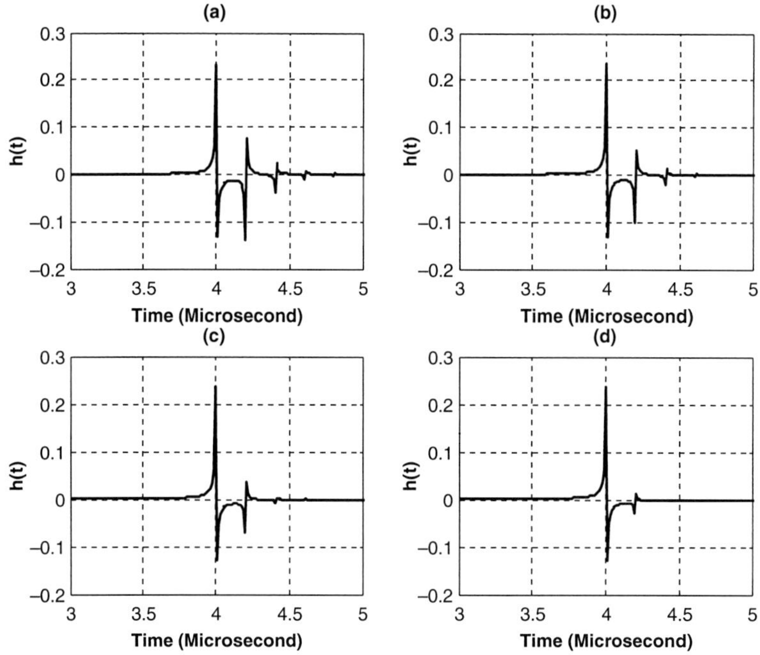

Figure 27: Impulse response for low voltage channel with a branch terminated in (a) 1 Ω (b) 100 Ω (c) 200 Ω (d) 400 Ω.

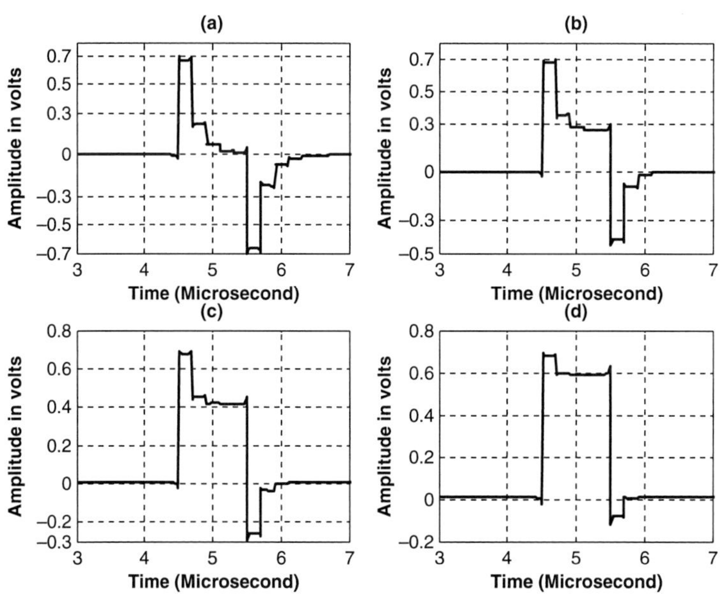

Figure 28: Received Signals at point C in Fig 1; for low voltage channel with a branch terminated in (a) 1 Ω (b) 100 Ω (c) 200 Ω (d) 400 Ω.

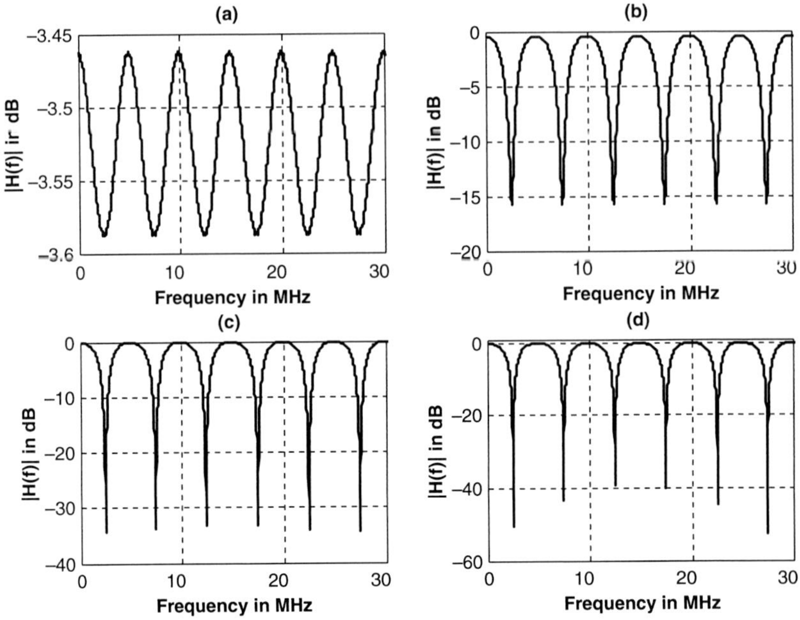

Figure 29: Transfer function response for low voltage channel with a branch terminated in (a) 600 Ω (b) 6 kΩ (c) 60 kΩ (d) 600 kΩ.

all the cases relating the voltage at the load and launched voltage at point A for different load conditions. It can be seen that as the load impedance tends to be higher than channel characteristic impedance, the peaks tend to improve and the notches tend to be deeper. The general expression for position of notches with open circuit load impedance is given by (7) and the corresponding position of peak frequency is given by (8).

Figure 30 shows the time domain impulse responses for all the four cases. It can be observed that the first paths are unchanged but as the load tends to open circuit the second and third paths tend to increase. Figure 31a–d shows the received signal for sending end signals of 2-V rectangular pulse, pulse width 1 μs, shifted by 0.5 μs. It can be observed that as the impedance tends to open circuit the received signal tends to be less attenuated.

$$f_i = \frac{75}{X}(1+2i), \quad \forall i = 0,1,2... \tag{7}$$

$$f_i = \frac{150}{X}i \quad \forall i = 0,1,2... \tag{8}$$

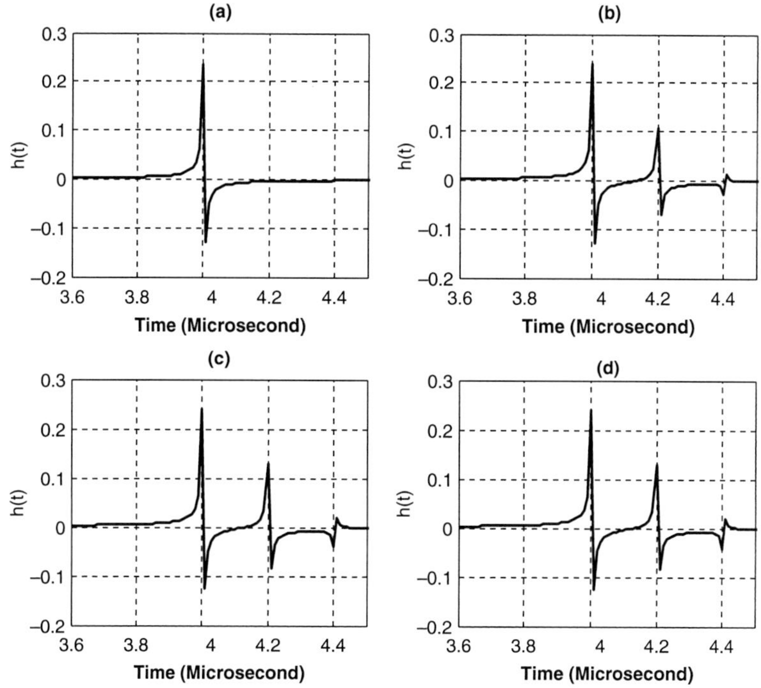

Figure 30: Impulse Response for low voltage channel with branch terminated in (a) 600 Ω (b) 6 kΩ (c) 60 kΩ (d) 600 kΩ.

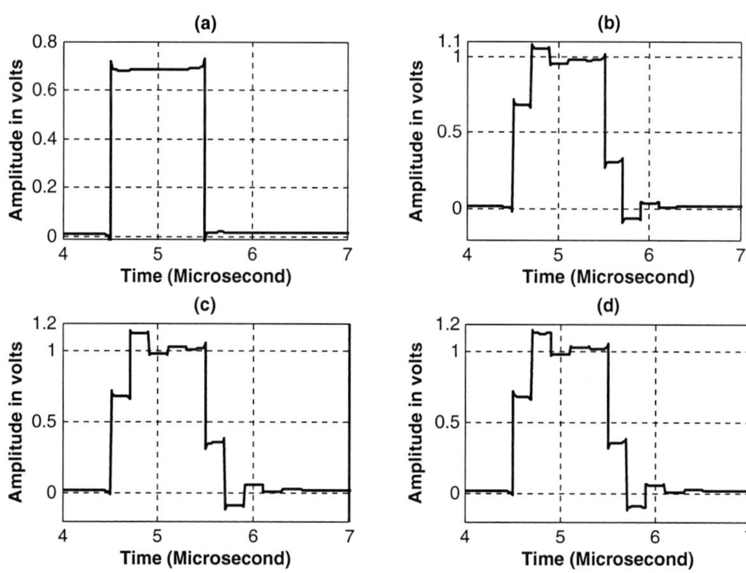

Figure 31: Received Signals at point C in Fig 1; for low voltage channel with branch terminated in (a) 600 Ω (b) 6 kΩ (c) 60 kΩ (d) 600 kΩ.

4 Indoor power-line channel

4.1 Effects of line length

The indoor power-line discussed here is similar to the systems used in Tanzanian residences or offices which connect appliance using cables line parameters L_e = 0.44388 μH/m, C_e = 61.734 pF/m [7]. The configuration under investigation is similar to power line shown in Fig. 1, with $Z_s = Z_L = 85\ \Omega$. The length AC was kept constant and equal to 20 m. The branch line length BD was varied as (BD = 5, 10, 15 and 20 m). Point D was terminated in 50 Ω. The transfer function was calculated by taking ratio of voltage at point C to the voltage at point A.

Figure 32a–d shows the result of the corresponding frequency responses for various branch line lengths. From Fig. 32, it can be seen that the peak values of signal responses were not attenuated significantly with either frequency or line length. The positions of peaks and notches are case dependent. The generalized expression for frequency position (f_i in MHz) of the ith peak in terms of branch line length (X in m) is approximately given by (9). Similarly the positions of the notches are given by (10). It was observed that the phase responses are less dependent on branch length.

$$f_i = \frac{47.8\ \text{MHz}}{X}(1+2i) \quad \forall\, i = 0,1,2... \tag{9}$$

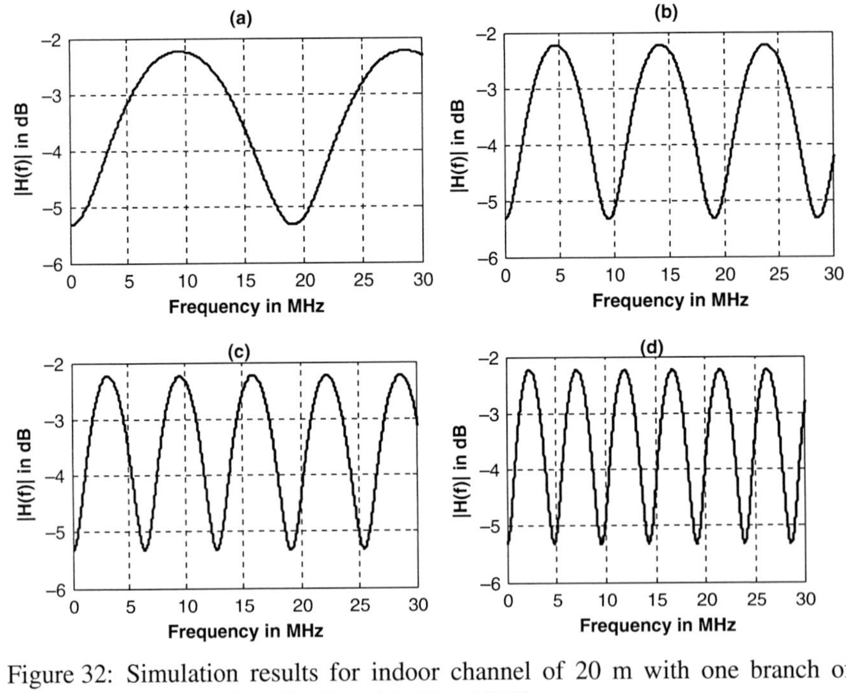

Figure 32: Simulation results for indoor channel of 20 m with one branch of length (a) 5 m (b) 10 m (c) 15 m (d) 20 m.

$$f_i = \frac{95.5 \text{ MHz}}{X}(i) \quad \forall i = 0,1,2... \tag{10}$$

4.2 Effects of number of branches

4.2.1 Multiple branches at single node

For this case the configuration as shown in Fig. 19 is considered. The length AB was fixed at 20 m and each branch length, e.g. BE was 10 m and terminated in 50 Ω. The number of branches was varied as 2 or 4 or 8 or 16. Figure 33a–d shows frequency responses for configuration with different number of branches.

It can be observed that the notches and peaks frequency positions are unchanged. But the increase in the number of branches at a single node leads to increased attenuation by approximately 2 dB/branch. The phase response was not significantly affected. Figure 34a–d shows the received pulses for an indoor channel with 2, 4, 8 and 16 branches at the same node for a 2-V rectangular input signal with pulse width of 1 μs, shifted by 0.5 μs. It can be observed that the peak values are

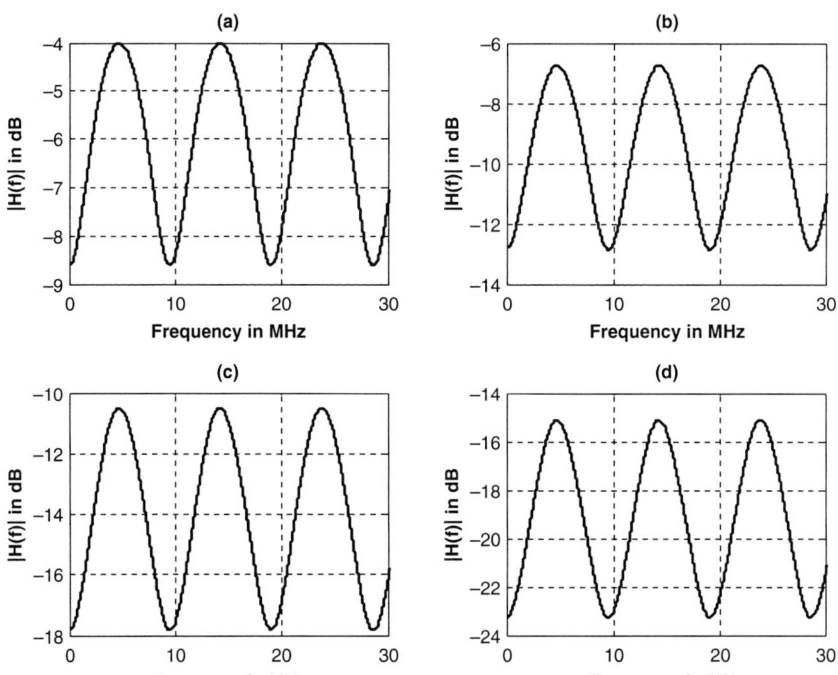

Figure 33: Simulation results for indoor channel with multiple branches at single node (a) 2 (b) 4 (c) 8 (c) 16 branches.

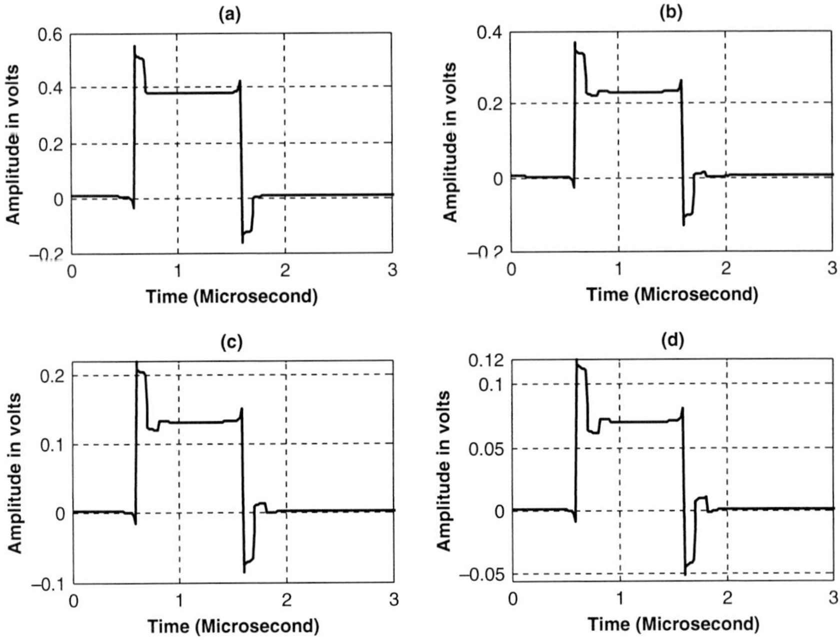

Figure 34: Received pulses at point C as in Fig 19; for indoor channel with multiple branches at single node (a) 2 (b) 4 (c) 8 (d) 16 branches.

attenuating as expected. Hence as the number of branches increases at the same node the signal suffers more attenuation.

4.2.2 Distributed branches

Another important case would be to study the effects of distributed branches as shown in Fig. 7. The length of AJ was kept at 20 m, while branches were 10 m with 50 Ω terminations as in earlier cases, but the number of distributed branches was varied as 2 or 4 or 8 branches. Note that the branches were equally distributed between AJ. Figure 35a–c, shows the frequency responses for different number of distributed branches. It is observed that the position of notches or peaks do not depend on the number of branches.

Figure 36a–d shows the corresponding phase responses. It is observed that as the number of branches increases the phase responses tend to lose linearity, e.g. compare Fig. 36a and c. This limits the available bandwidth and the channel has distortions. Figure 37 shows the received signals for a transmitted rectangular pulse of 2 V, pulse width of 1 μs, shifted by 0.5 μs. As the number of distributed branches increases the signals suffer both attenuations and distortion. The frequency responses could be also affected by the termination impedances which are discussed next.

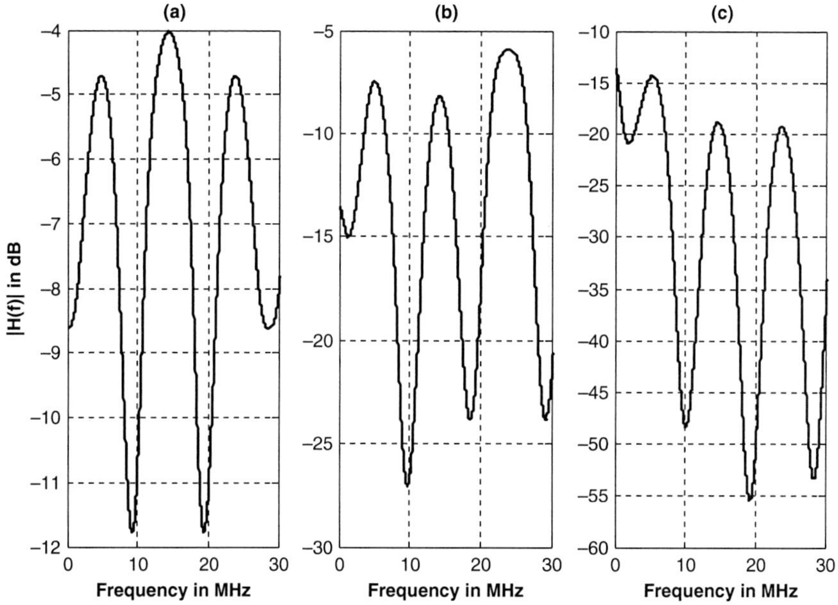

Figure 35: Simulated frequency response results for indoor channel with distributed branches (a) 2 (b) 4 (c) 8 branches.

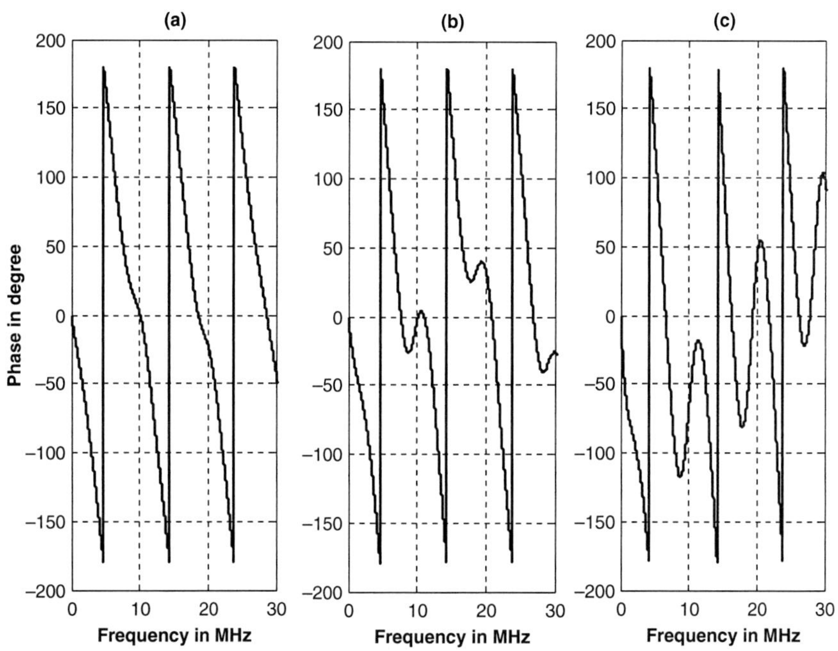

Figure 36: Simulated phase response results for Indoor channel with distributed branches (a) 2 (b) 4 (c) 8 branches.

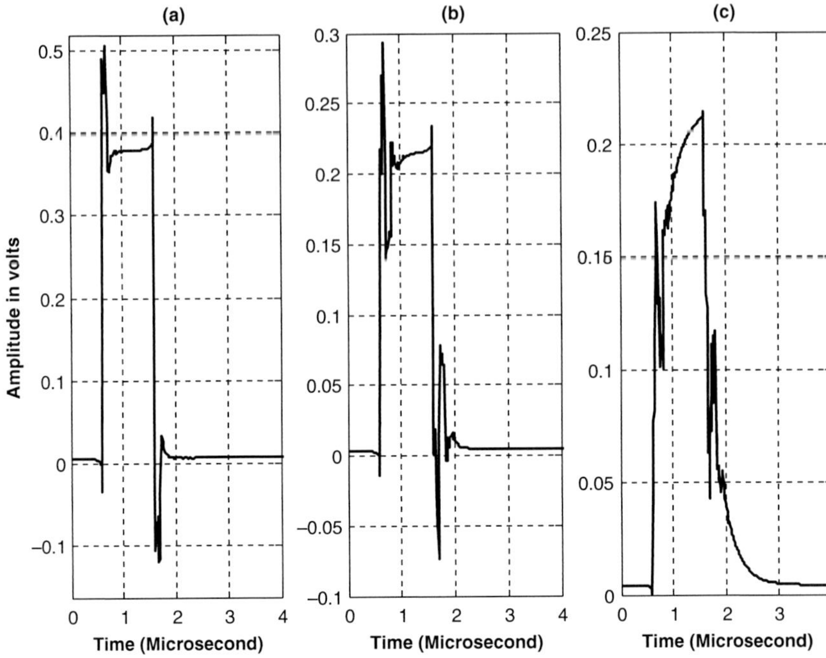

Figure 37: Received signals at point J as in Fig 7; for indoor channel with distributed branches (a) 2 (b) 4 (c) 8 branches.

4.3 Effects of load

In this study the terminal impedance at the branches were divided into two different cases. The first case is resistive load which is varied between 1 Ω and 1 kΩ is considered in accordance with the measurements of Neto *et al.* [8]. Secondly varying inductive load is considered (RL loads, varied between 0.1 and 100 mH, with constant resistance of 2 Ω, in accordance with Keyhani & Birtwhistle [9]. In all investigations Fig. 1 was considered and the impedance at terminal D was varied. The length AC was 20 m while the length of BD was 10 m.

In this investigation the impedances were varied as 4 Ω, 40 Ω, 85 Ω, 400 Ω, 800 Ω and 4 kΩ. Figure 38a–d shows the corresponding frequency responses. For load impedance characteristics impedance less than line characteristic impedance the positions of peaks and notches are unchanged. It is observed that as impedance approaches short circuit the notches attenuation increase and the same behavior is observed when it approaches open circuit compared to line characteristic impedance. Figure 39a–d shows the corresponding phase responses for indoor power-line channel for various termination impedances.

It is observed that as impedances approach short circuit the phase tends to be more distorted compared to a situation when load impedances approaches characteristic impedance, while an improvement in the linearity of phase responses is

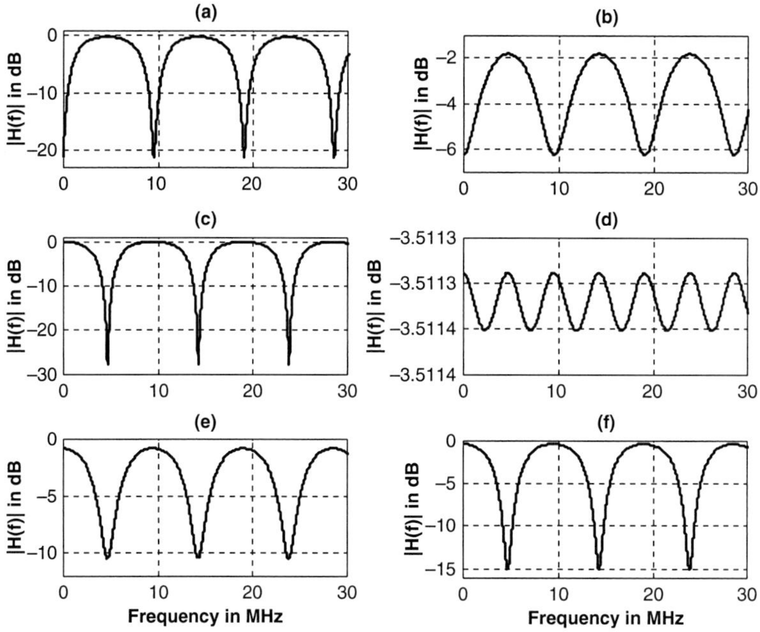

Figure 38: Results for an indoor channel with a branch terminated in low impedances (a) 4 Ω (b) 40 Ω (c) 85 Ω (d) 400 Ω (e) 800 Ω (f) 4 kΩ.

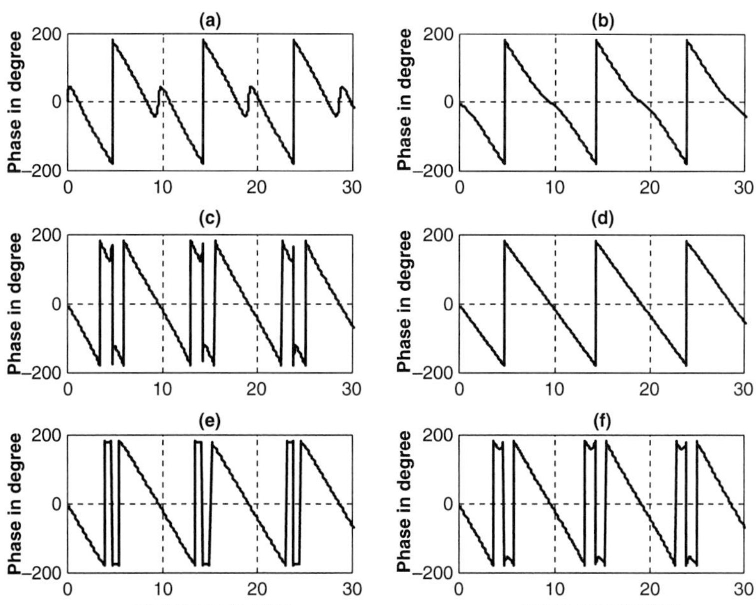

Figure 39: Phase response for channel with a branch terminated in (a) 4 Ω (b) 40 Ω (c) 85 Ω (d) 400 Ω (e) 800 Ω (f) 4 kΩ.

seen. Thus, as the termination impedances are much lower than characteristic impedance the available bandwidth is limited. Figure 40a–d shows the received signals for a 1-V rectangular signal with pulse width of 1 μs and shifted by 0.5 μs. It is observed that as the impedances tend to lower values the signal experiences both attenuations and distortions compared to characteristic impedance case where only attenuations are observed. Further only attenuations are observed as the loads are greater than characteristic impedance. The peaks in the frequency response occur at a frequency given by (11); the parameter X is the length of branch cable. Similarly, the notches occur at a frequency as given by (12). Expressions (11) and (12) are approximate and applicable when the load impedance is more than characteristic impedance.

$$f_i = \frac{95.5\text{ MHz}}{X}(i) \quad \forall i = 0,1,2... \tag{11}$$

$$f_i = \frac{47.8\text{ MHz}}{X}(1+2i) \quad \forall i = 0,1,2... \tag{12}$$

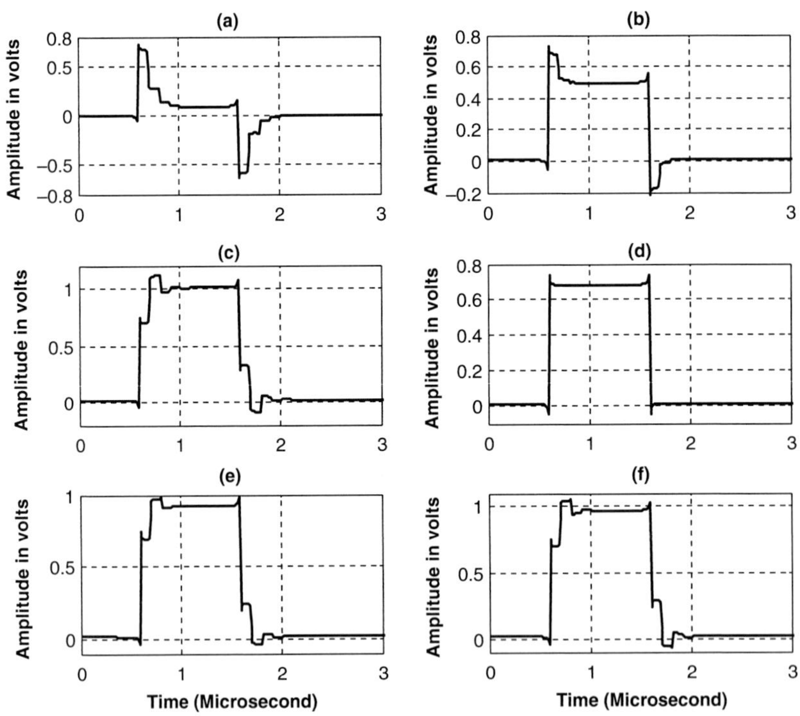

Figure 40: Received pulses at point C as in Fig 1; for indoor channel with a branch terminated in (a) 4 Ω (b) 40 Ω (c) 85 Ω (d) 400 Ω (e) 800 Ω (f) 4 kΩ.

In the investigations with the inductive loads which were varied as 0.1, 1, 10 and 100 mH with a constant resistance of 2 Ω the frequency responses were similar to Fig. 38f with an exception that the notch attenuations of –30 dB were observed. Thus the inductive loads act as high impedance loads.

5 Using infinite return ground in BPLC systems – transmission line analysis

5.1 Transmission lines with return ground

The case of a single overhead power-line with ground as a return conductor is shown in Fig. 41. The finitely conducting ground is characterized by its conductivity (σ_g), permittivity ($\varepsilon_g = \varepsilon_{rg}\varepsilon_0$) and free space permeability (μ_0), the parameters ε_{rg} and ε_0 are relative ground permittivity and permittivity in free space, respectively. For ground return systems, per unit impedance and admittance are discussed in Chapter 2.

5.2 Influence of signal propagation from transmitter to receiver

A typical medium voltage line of Tanzanian power network is considered. First, a case of transmission line with adjacent return with source impedance and load impedance $Z_L = Z_S = 624\ \Omega$ is considered without any interconnection. The line

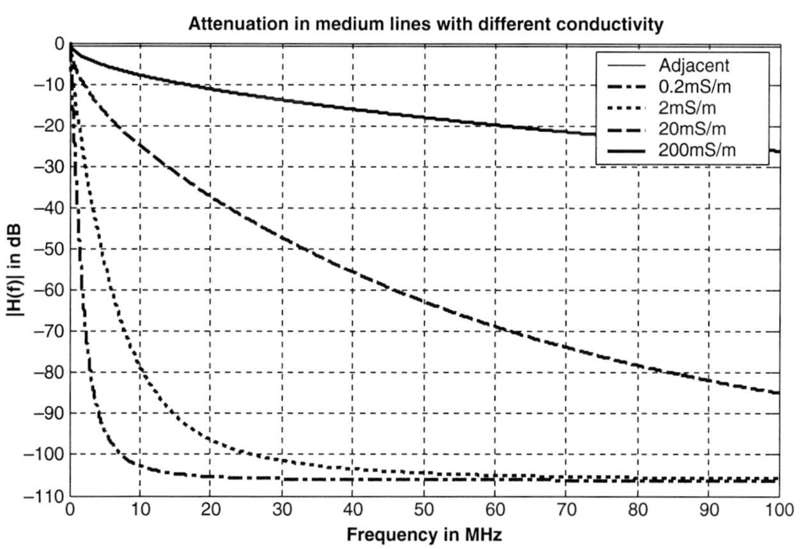

Figure 41: Frequency response simulation results for medium voltage lines for a direct length from transmitter to receiver for different ground conductivity with fixed ground relative permittivity of 10.

length is 4 km which is a typical line length in MV line. For the adjacent conductor return the conductor separation was 1 m and conductor radius, 100 mm^2. The line has the per unit length parameters with $L_e = 1.9648$ μH/m, $C_e = 5.6627$ pF/m and $R = 0.1472$ mΩ/m [10]. For the case with ground return the conductor height $h_k = 11$ m above ground. The ground conductivity (σ_g) considered were 200, 20, 2 and 0.2 mS/m; ground relative permittivity of 10 was constant for all cases.

The simulation results for the transfer function are shown in Fig. 41. In order to study the influence of ground relative permittivity, simulations were carried out for the case with ground return with ground relative permittivity varied as 2, 4, 8 and 16 and keeping the ground conductivity at 20 mS/m. Simulation results for the transfer function are shown in Fig. 42. Note the transfer functions are relating the load voltages and the launched voltages at sending end. Frequency responses of up to 100 MHz were considered due to proposed frequency bands for medium voltage. From Fig. 41, it can be observed that as ground conductivity tends to decrease attenuations tends to increase with frequencies. For adjacent return, the attenuation is more or less unaffected and is around 2 dB compared to 100 dB for ground with low conductivity.

From Fig. 42, it is seen that as the frequencies increase for a ground with lower permittivity, it can lead to increased attenuation beyond 10 dB for frequencies above 60 MHz.

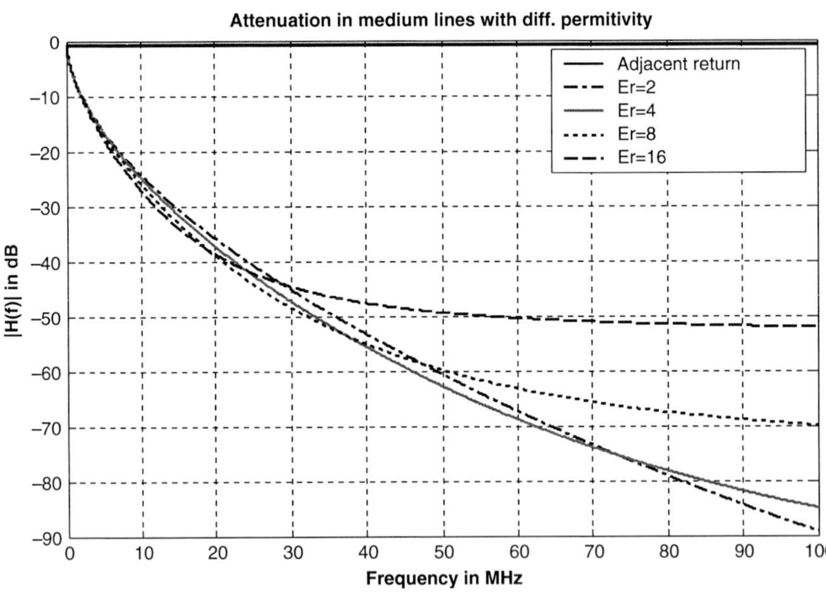

Figure 42: Frequency response simulation results for medium voltage lines for a direct length from transmitter to receiver for different ground relative permittivities.

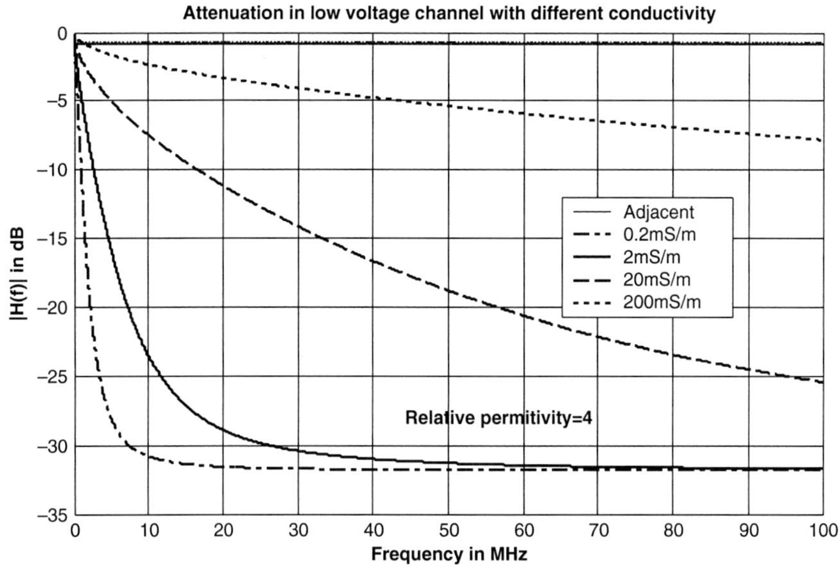

Figure 43: Frequency response simulation results for low voltage lines for a direct length from transmitter to receiver for different ground conductivity.

A typical LV network that has line parameters L_e = 1.9589 μH/m, C_e = 5.6799 pF/m and R = 0.1472 mΩ/m was considered. The configuration was as shown for MV network but with $Z_L = Z_s$ = 587 Ω with line length 1.2 km. For the ground return case the conductor height h_k = 10 m above ground. The ground conductivity (σ_g) was considered as 200, 20, 2 and 0.2 mS/m, ground relative permittivity of 10 as in the previous case; then ground conductivity was kept constant while varying ground relative permittivity as 2, 4, 8 and 16. Figures 43 and 44 show frequency responses of transfer functions relating the load voltages and the launched voltage at sending end for the varying ground conductivity case and varying ground relative permittivity case, respectively. From Fig. 45, it can be observed that as ground conductivity decreases attenuations increase with frequency but less severe compared to the medium voltage line (compare Fig. 43). For adjacent return the attenuation is in the range of 5 dB. From Fig. 46, it is seen that as the frequency increases, the relative permittivity does not influence the channel attenuation in the frequency band under consideration compared to the medium voltage line (compare Fig. 44).

5.3 Influence of signal propagation with respect to branched line length

The configuration as in Fig. 1 was considered; i.e. the length of the line from point A to C was kept constant at 4 km. The branched length was varied as BD = 10, 20, 40 and 80 m with B always at the midpoint of line AC. Point D was terminated in 50 Ω, and the transfer characteristic was calculated with respect to the load at C similar to earlier cases, i.e. one with adjacent conductor return and the other with

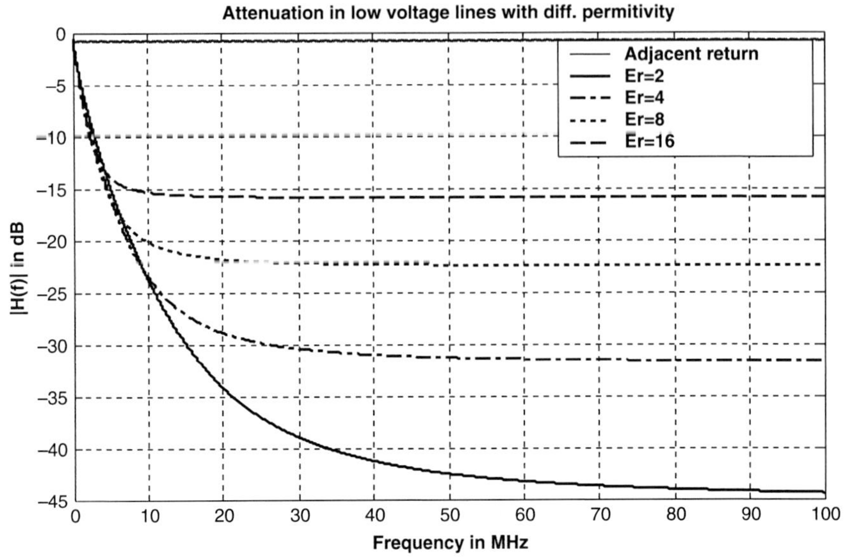

Figure 44: Frequency response simulation results for low voltage lines for a direct length from transmitter to receiver for different ground relative permittivities.

ground as return. The ground conductivity (σ_g) was considered as 20 mS/m and the ground relative permittivity of 10.

Figure 45a–d shows the corresponding frequency responses (ground and adjacent return) for various branch line lengths. It is observed in Fig. 46 that the number of peaks and notches for a given frequency window is increasing with branch length for either of the cases. For the case with ground return the frequency position of the either the notch or the peak, i.e. the attenuation is increasing linearly with frequency. Next the case of low voltage network was considered as in Fig. 1 with length between A and C kept constant at 1.2 km. The branched length was varied as BD = 10, 20, 40 and 80 m with B always at the mid of line AC. Point D was terminated in 50 Ω, and the transfer characteristic was calculated with respect to the load at C for the case of adjacent conductor return and the ground return. The ground conductivity (σ_g) was considered as 20 mS/m and the ground relative permittivity of 10 similar to the previous case.

Figure 47a–d shows the corresponding frequency responses for various branch line lengths. It is observed that similar to the previous case the number of peaks and notches for a given frequency window increases with line length with either the adjacent conductor return or ground return. For the case with ground return the frequency position of either the notch or the peak, i.e. the attenuation increases linearly with frequency. Compared to the medium voltage case as shown in Fig. 45, it can be seen that the number of peaks and notches for a given frequency window in low voltage channel is less; at the same time the attenuation is also less compared to the medium voltage case.

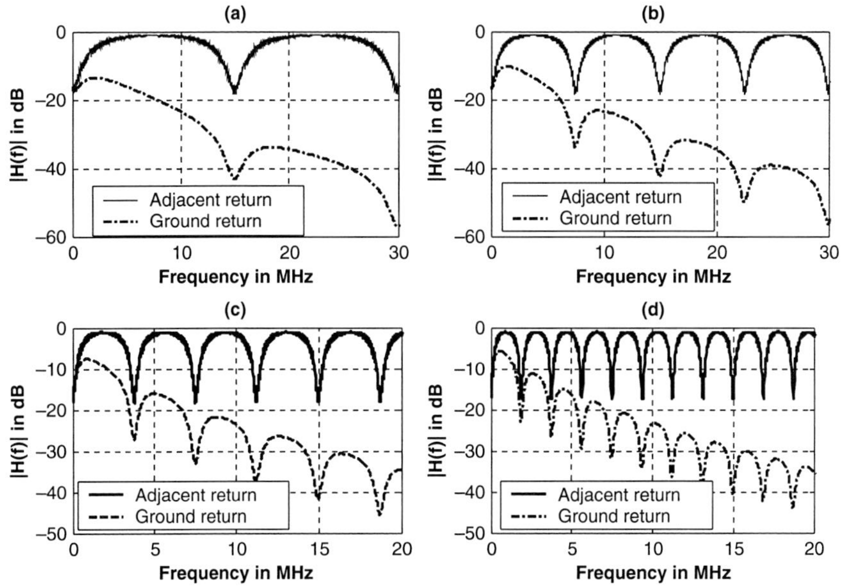

Figure 45: Frequency response of medium voltage lines for a power-line network with branched length of (a) 10 m (b) 20 m (c) 40m (d) 80m.

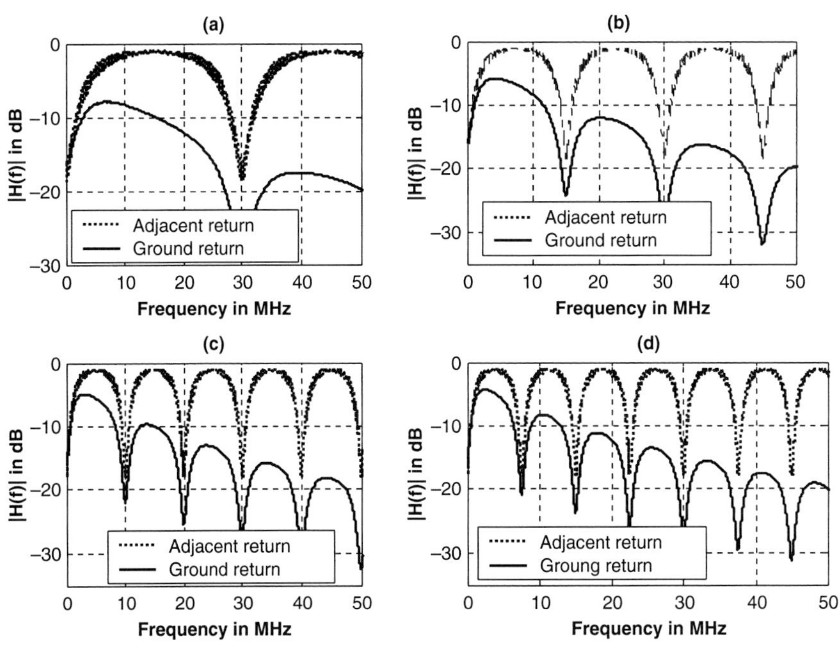

Figure 46: Frequency response of low voltage lines for a power-line network with branched length of (a) 5 m (b) 10 m (c) 15 m (d) 20 m.

6 Underground cables for BPLC systems: frequency response

6.1 Influence of line length

6.1.1 Influence of length from transmitter to receiver

The network configuration is shown in Fig. 1 with $Z_L = Z_s = Z_c$, the characteristic impedance of the line ABC. The low voltage power-line cables most used for underground network are considered, whereby the line ABC is NAYY150SE branched with BD which is NAYY35RE. With line parameters as $L_1 = 0.32735$ μH/m, $C_1 = 0.\ 27191$ pF/m (for section ABC) and $L_2 = 0.45179$ μH/m, $C_2 = 0.19702$ pF/m (section BD) for NAYY150SE and NAYY35RE, respectively [11]. The line length AC was varied as 1.2 km, 600 m, 300 m and 150 m, with point B always at mid-point of AC. The branch line length (BD) was considered constant equal to 15 m and terminated in 10 kΩ. Figure 47a–d shows the transfer functions relating the voltages at the load and sending end for various lengths of AC. From Fig. 47, the peaks and notches in frequency response do not vary with either frequency or line length. The positions of peaks and notches are independent of the line length from transmitter to receiver. In addition, it can be observed that as the length of the line increases the attenuation increases.

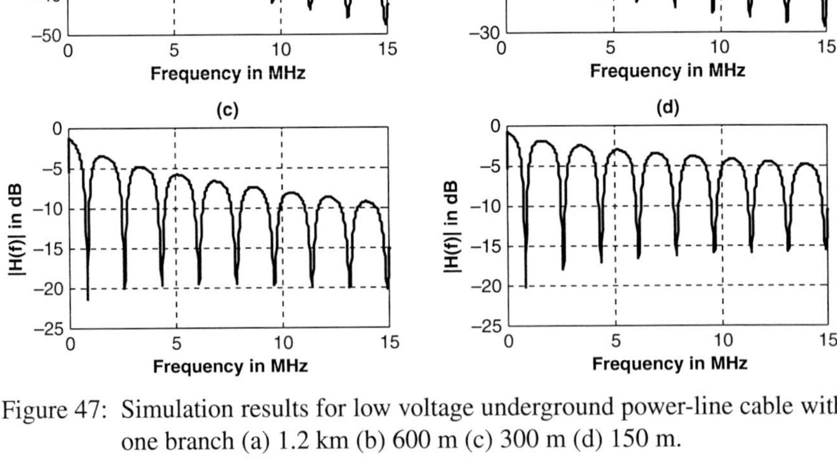

Figure 47: Simulation results for low voltage underground power-line cable with one branch (a) 1.2 km (b) 600 m (c) 300 m (d) 150 m.

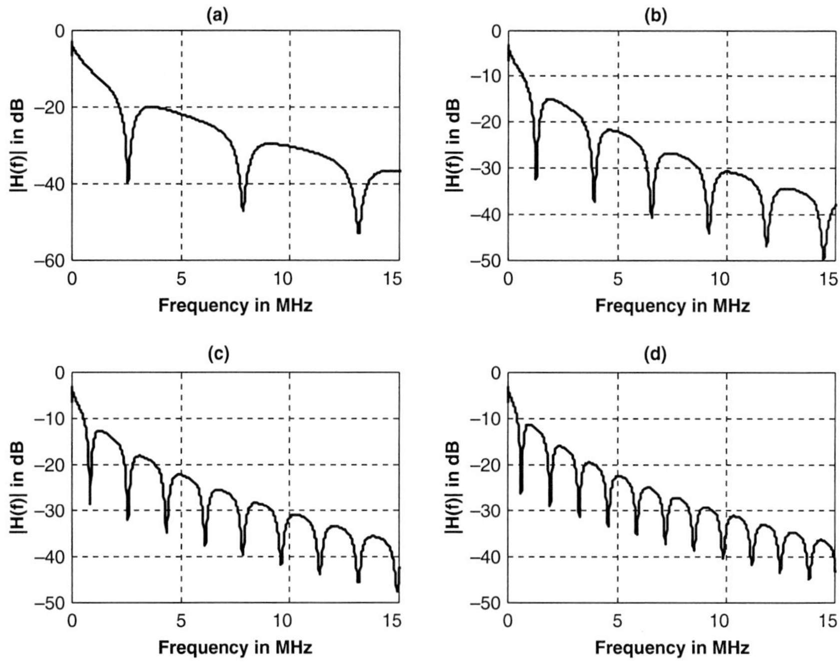

Figure 48: Simulation results for 1.2 km underground cable with one branch of length (a) 10 m (b) 20 m (c) 30 m (d) 40 m.

6.1.2 Influence of branch length

The configuration is same as in previous section. The length of AC was kept constant at 1.2 km, while the length of BD was varied as 10, 20, 30 and 40 m. Point D was terminated in 10 kΩ. The transfer functions for all cases relating the voltage at the load and launched voltage at point A are shown in Figure 48a–d. It can be observed that the positions of notches and peaks are dependent on the length of the branch line. As the branch line length increases, it results into more notches and peaks. The attenuation in each case increases with the frequency increase. The generalized expression for frequency position (f_i in MHz) of the ith notch in terms of branch line length (X in m) is approximately given by (13).

$$f_i = \frac{26}{X}(1+2.i), \quad \forall i = 0,1,2... \tag{13}$$

6.2 Influence of number of branches

6.2.1 Multiple branches at single node

Consider the configuration as in Fig. 19. The line AC is 1.2 km with all branches 15 m long concentrated at node B, with direct line length ABC cable being same

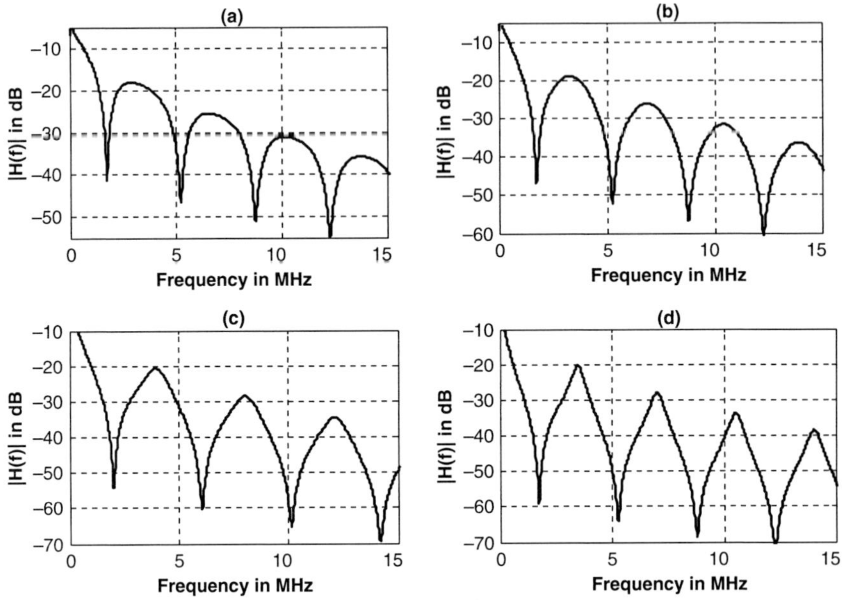

Figure 49: Simulation results for low voltage underground channel with multiple branches at single node (a) 2 (b) 4 (c) 8 (d) 16 branches.

as in previous section but the branched cables being different as discussed in previous section. All branches were terminated in 10 kΩ. The number of branches was varied as 2, 4, 8 and 16. Figure 49a–d shows the transfer functions for all cases relating the voltage at the load and launched voltage at point A. It can be observed that the notches attenuations decrease with increase in the number of branches.

6.2.2 Distributed branches

A low voltage underground channel with distributed branches as in Fig. 7 is considered. The number of branches was increased in the link between points A and J. The length AJ was 1.2 km while all branches were 15 m, with direct line length AJ cable being same as in previous section but the branched cables being different as discussed in previous section. The branches were varied as 2, 4 and 8 such that in each case they were equally distributed along AJ. The terminations were considered as 10 kΩ. Figure 50a–d shows the transfer functions for all cases relating the voltage at the load and launched voltage at point A, for different number of branches. It is seen that the positions of deep notches are not changed. As the number of branches increase the attenuations of notched point tend to increase. The stochastic attenuations are observed as the number of branches increases which increases the possibilities of reducing the available bandwidth for low voltage channel.

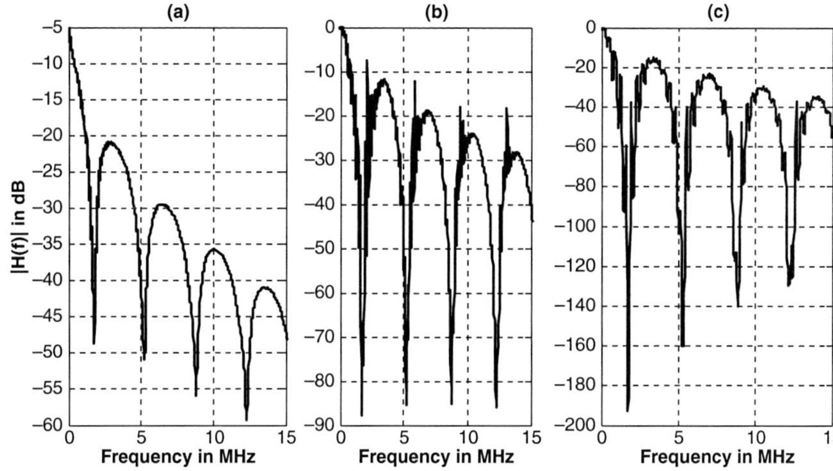

Figure 50: Simulation results for low voltage underground channel with distributed branches (a) 2 (b) 4 (c) 8 branches.

6.3 Influence of load impedance

6.3.1 Low resistive load

The effect of load impedances is considered. The configuration is as shown in Fig. 1, with cable configurations for direct line branched line as before. The load impedances at point D were varied as 5 Ω, 10 Ω, 20 Ω and Z_0 charactristic impedance. The length of AC was 1.2 km and BD was 15 m. Figure 51a–d shows transfer functions for all cases relating the voltage at the load and launched voltage at point A for different load conditions. It can be seen that as the load impedance increases towards characteristic impedance, the peaks decrease and the notches improve.

6.3.2 High resistive load

Next considering the same case as before higher impedances for the load was considered with the variation as 50 Ω, 100 Ω, 1 kΩ and 10 kΩ. Figure 52a–d shows transfer functions for all cases relating the voltage at the load and launched voltage at point A for different load conditions. It can be seen that as the load impedance becomes higher than channel characteristic impedance, the peaks improve and the notches become deeper. The general expression for position of notches with open circuit load impedance is given by (14).

$$f_i = \frac{52}{X}(1+2i), \quad \forall i = 0,1,2... \tag{14}$$

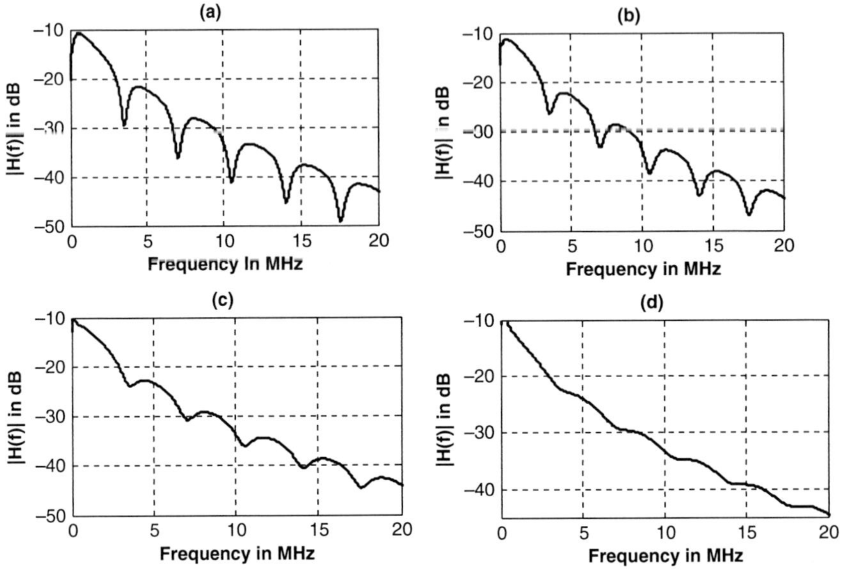

Figure 51: Simulation results for low voltage underground channel with distributed branches (a) 5 Ω (b) 10 Ω (c) 20 Ω (d) Z_o.

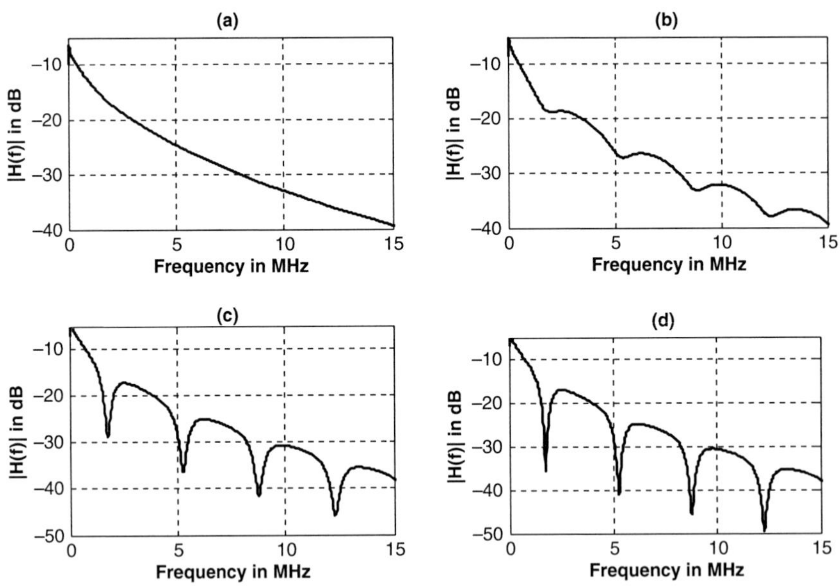

Figure 52: Transfer function response for low voltage underground channel with distributed branches (a)50 Ω (b) 100 Ω (c) 1 kΩ (d) 10 kΩ.

References

[1] Mathias, G., Rapp, M. & Dostert, K., Power line channel characteristics and their effect on communication system design. *IEEE Communications Magazine*, pp. 78–86, April 2004.

[2] Pavlidou, N., Vinck, A.J.H., Yazdani, J. & Honary, B., Power line communications: state of the art and future trends. *IEEE Communications Magazine*, pp. 34–40, April 2003.

[3] Zimmermann, M. & Dostert, K., A multipath model for the powerline channel. *IEEE Transactions on Communications*, **50(4)**, pp. 553–559, April 2002.

[4] Anatory, J., Theethayi, N. & Thottappillil, R., Power-line communication channel model for interconnected networks – part I: two conductor system. *IEEE Transactions on Power Delivery*, **24(1)**, January 2009.

[5] Anatory, J., Theethayi, N., Thottappillil, R., Kissaka, M. M., & Mvungi, N. H., The Effects of Load Impedance, Line Length and Branches in the BPLC-Transmission Lines Analysis for Medium Voltage Channel. *IEEE Transactions on Power Delivery*, **22(4)**, pp. 2156–2162, October 2007.

[6] Anatory, J., Theethayi, N., Thottappillil, R., Kissaka, M.M. & Mvungi, N.H., The effects of load impedance, line length and branches in the BPLC based on transmission lines analyses for typical a low voltage channel in developing countries. *IEEE Transactions on Power Delivery*, **24(4)**, April 2009.

[7] Anatory, J., Theethayi, N., Thottappillil, R., Kissaka, M.M. & Mvungi, N.H., The effects of load impedance, line length and branches in the BPLC – transmission lines analysis for indoor voltage channel. *IEEE Transactions on Power Delivery*, **22(4)**, pp. 2150–2155, October 2007.

[8] Neto, J.P., Tsuzuki, S., Kawakami, Y. & Yamada, Y., In-door powerline impedance measurements up to high frequency (10kHz–70MHz). *Proceedings of 7th ISPLC03*, Kyoto, Japan, pp. 243–247, 2003.

[9] Keyhani, R. & Birtwhistle, D., Modeling High Frequency Signal Propagation over Low Voltage Distribution Lines. *Australasian Universities Power Engineering Conference (AUPEC 04)*, September 2004.

[10] Anatory, J. & Theethayi, N., On the efficacy of using ground return in the broadband power line communications – a transmission line analysis. *IEEE Transactions on Power Delivery*, **23(1)**, pp. 132–139, January 2008.

[11] Anatory, J., Theethayi, N., Thottappillil, R., Kissaka, M.M. & Mvungi, N.H. The influence of load impedance, line length and branches on underground cable power-line communications (PLC) systems. *IEEE Transactions on Power Delivery*, **23(1)**, pp.180–187, January 2008.

CHAPTER 5

Channel characterization for different PLC systems

1 Introduction

Researchers in the recent past have been looking into determining channel parameters for improving the channel performance. Much of those studies were focusing on frequencies up to 30 MHz [1–5]. However, using this frequency range it has been observed that the attainable data rate for indoor power-line is 200 Mbps and with effective data rate 70 Mbps [6, 7]. To be able to attain good quality of services, service frequencies up to 100 MHz is being proposed [8, 9]. In this frequency band the knowledge of channel characterization and signal propagation is needed. There has been limited knowledge on signal propagation aspects from a theoretical angle, and attempts are directed towards measurements [10]. One of the reasons which limit the investigations could be the complexity of channel models to determine different impulse responses channel models in the proposed frequency bands. There are some factors which need to be well researched for proper performance improvements. These factors are channel delay spread and channel capacity. Delay spread is necessary for determining limits on data rates due to channel inter-symbol interference. This can lead to improvements of modulation scheme designs such as orthogonal frequency division multiplexing and direct sequence-code division multiple access systems [9].

For example power-line network consists of terminal loads, branches and different line lengths. To be able to characterize a power line theoretically one has to determine [11–13]:

- To what extent the number of branches affects the delay spread in the given frequency bands?
- How the terminal loads (infinite and low) and impedances affect the channel delay spread?

In this chapter different channel impulse responses for medium voltage (MV), low voltage (LV) and indoor voltage (IV) power-line network have been investigated. The investigations look at the network with different branches such as 4, 8 and

12 branches. In each case the analysis of the branch impedances were terminated in higher impedances, low impedances and characteristics impedances. In all cases the channel impulse responses were determined. Since a power-line network exhibits a time/frequency variation, the responses were averaged to get the appropriate channel impulse response [14–16].

2 Analysis of channel delay parameters

In order to compare different multipath channels and develop some general design guidelines for broad band power line communication (BPLC) systems, parameters which grossly quantify the multipath channel must be used [11]. The mean excess delay, rms delay spread and excess delay spread (x dB) are multipath channel parameters that can be determined from power delay profile. The mean excess delay ($\overline{\tau}$) and root mean square (rms) delay spread (σ_τ) are given by (1) and (2), respectively, where ($\overline{\tau^2}$) is given by (3). The maximum excess delay τ_{m} is a time measured with respect to a specific power level, which is characterized as the threshold of the signal. When the signal level is lower than the threshold, it is processed as noise [17, 18].

$$\overline{\tau} = \frac{\sum_k a_k^2 \tau_k}{\sum_k a_k^2} = \frac{\sum_k P(\tau_k)\tau_k}{\sum_k P(\tau_k)} \tag{1}$$

$$\sigma_\tau = \sqrt{\overline{\tau^2} - (\overline{\tau})^2} \tag{2}$$

$$\overline{\tau^2} = \frac{\sum_k a_k^2 \tau_k^2}{\sum_k a_k^2} = \frac{\sum_k P(\tau_k)\tau_k^2}{\sum_k P(\tau_k)} \tag{3}$$

$$P(\tau) = \left|h(t;\tau)\right|^2 \tag{4}$$

3 Analysis of coherence bandwidth parameters

Coherence bandwidth is a statistical measurement of the range of frequencies over which the channel can be considered flat or in other words the approximate maximum bandwidth or frequency interval over which two frequencies of a signal are likely to experience comparable or correlated amplitude fading. If the multipath rms delay spread (σ_τ) seconds, then the coherence bandwidth B_{C} over which the frequency correlation function (Δf) is 0.9 and 0.5 is approximated by (5) and (6), respectively [18].

$$B_{\mathrm{C}} \approx \frac{1}{50\sigma_\tau} \tag{5}$$

$$B_C \approx \frac{1}{5\sigma_\tau} \tag{6}$$

4 Analysis of channel capacity

The channel capacity of an ideal, band-limited, AWGN channel is given by (7), where C is the capacity in bits/s, W is the channel bandwidth and P_{av} is the average transmitted power. In a multi-carrier system, with (Δf) sufficiently small, the sub-channel has a capacity as in (8). The total capacity of the channel is given by (9). The parameters $\Phi_{nn}(f_i)$ and $P(f_i)$ are power spectral density of the noise and power signal of the transmitted signal, respectively [17, 18]. The channel transfer function calculations is discussed in Chapter 3.

$$C = W\log_2\left(1+\frac{P_{av}}{WN_o}\right) \tag{7}$$

$$C_i = \Delta f\log_2\left(1+\frac{\Delta f P(f_i)|H(f_i)|^2}{\Delta f\Phi_{nn}(f_i)}\right) \tag{8}$$

$$C = \Delta f\sum_{i=1}^{N}\log_2\left(1+\frac{P(f_i)|H(f_i)|^2}{\Phi_{nn}(f_i)}\right) \tag{9}$$

5 Characterization of different PLC systems

5.1 Medium voltage systems

The MV power channel similar to the systems in Tanzanian power line network is considered for study. The line parameters are $L_e = 1.9648$ μH/m, $C_e = 5.6627$ pF/m [16, 19, 20]. The investigations considered the network with 4, 8 and 12 branches. The configuration under study is as in Fig. 1.

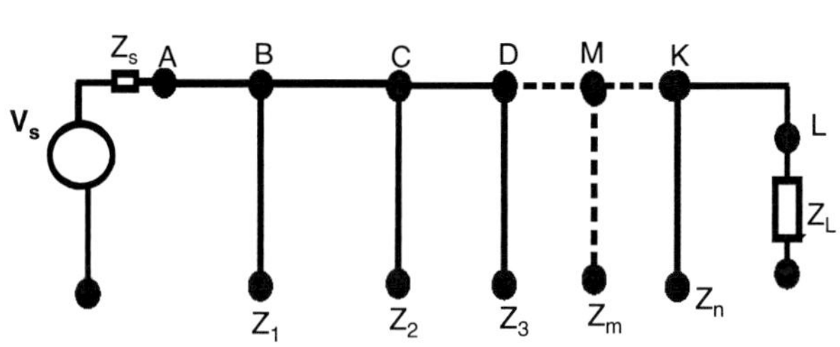

Figure 1: Power-line network with distributed branches.

5.1.1 Channel with four distributed branches

Consider the configuration as in Fig. 1 with $Z_s = Z_L = 589\ \Omega$. The number of distributed branches between point A and L is four. The length A–L was kept constant and equal to 4 km. The branch line lengths were kept at 30 m. The terminal loads at distributed branches were varied as terminated in (50 Ω, characteristic impedance and 10 kΩ). Note that the branches were equally distributed between A and L.

For each case the important channel parameters were investigated. Figure 2 shows the multipath power delay profiles of a power-line link with four distributed branches. Figure 3 shows the delay profiles of power-line link with four branches and terminated in various loads impedances. Figure 3d is averaged delay profile which is obtained after averaging the responses of Fig. 3a–c. It should be noted that the attenuations of multipaths increase as the channel is terminated in characteristic impedances. From Fig. 3, the channel characterization parameters were calculated and are as indicated in Table 1. The threshold of –30 dB from the maximum value was considered; the maximum excess delay τ_m is 11 μs for all cases of impedances. Table 1 shows channel parameters when the load terminals are terminated in high impedances, characteristic impedances, lower impedances and averaged impulse response, respectively. It can be observed that the averaged delay

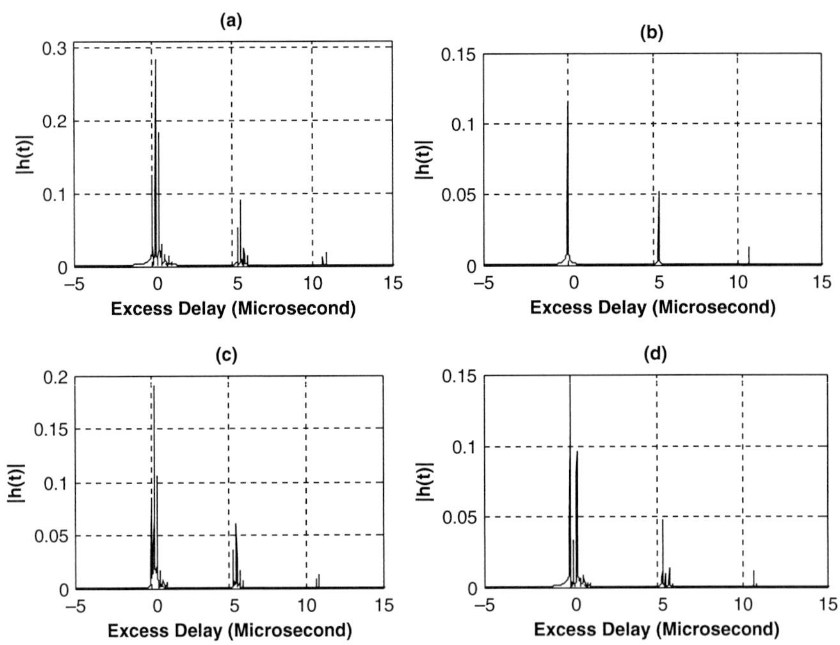

Figure 2: Simulated multipath impulse response for medium voltage power line link with four distributed branches (a) all terminated in 10 kΩ (b) all terminated in characteristic impedances (c) all terminated in 50 Ω (d) averaged response.

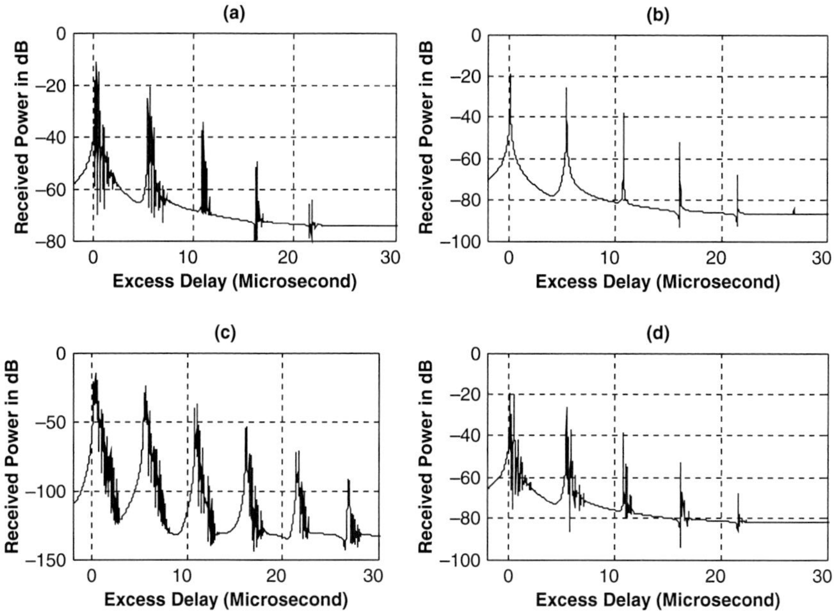

Figure 3: Simulated multipath power delay profiles for medium voltage power-line link with four distributed branches (a) all terminated in 10 kΩ (b) all terminated in characteristic impedances (c) all terminated in 50 Ω (d) averaged delay profile.

Table 1: Channel parameters for an MV network with four branches –30dB, 0.1–100 MHz.

Branch loads	τ (μs)	σ_τ (μs)	$B_{C-90\%}$ (kHz)	$B_{C-50\%}$ (MHz)
10 kΩ	0.7146	1.6005	12.496	0.12496
50 Ω	0.7413	1.6550	12.085	0.12085
Char. impedance	0.9964	2.2233	8.9955	0.089955
Average	0.4939	1.4560	13.736	0.13736

spread for a channel with four branches is 1.4560 μs; this corresponds with 90% coherence bandwidth of 13.736 kHz.

5.1.2 Channel with eight distributed branches

Consider the configuration as in Fig. 1 with $Z_s = Z_L = 589\ \Omega$. The number of distributed branches between point A and L is eight. The length A–L was kept constant and equal to 4 km. The branch line length was kept at 30 m. The terminal loads at distributed branches were varied as terminated in (50 Ω, characteristic

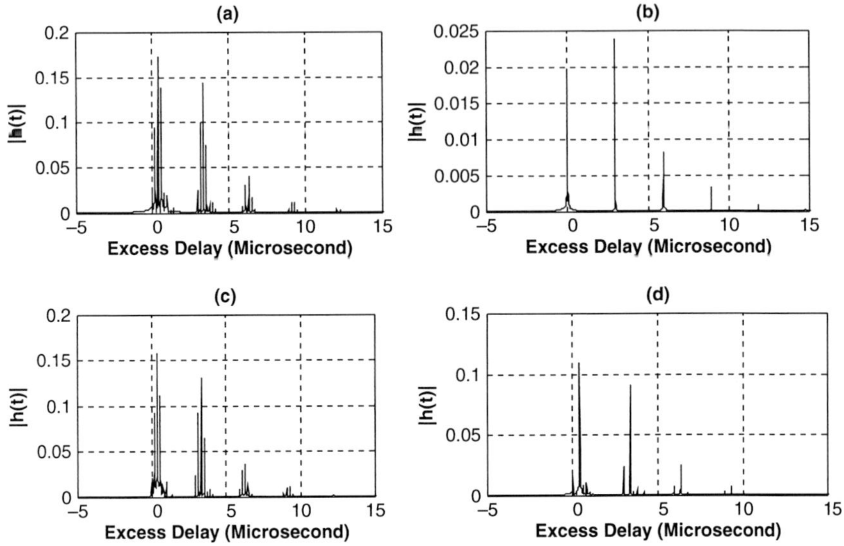

Figure 4: Simulated multipath impulse response profiles for medium voltage power-line link with eight distributed branches (a) all terminated in 10 kΩ (b) all terminated in characteristic impedances (c) all terminated in 50 Ω (d) averaged delay profile.

impedance and 10 kΩ). Note that the branches were equally distributed between A and L. For each case the important channel parameters were investigated. Figure 4 shows the multipath channel impulse response of a power-line link with eight distributed branches. Figure 5 shows the delay profiles of power-line link with eight branches and terminated in various loads impedances. Figure 5d is averaged delay profile which is obtained after averaging the responses of Fig. 5a–c. It should be noted that the attenuations of multipaths increase as the channel is terminated in characteristic impedances. From Fig. 5, the channel characterization parameters were calculated and are as indicated in Table 2.

The threshold of –30 dB from the maximum value was considered; the maximum excess delay τ_m is varying between 10, 11.9, 10 and 9.3 μs for a channel terminated in high impedances, characteristic impedances, low impedances and averaged channel response, respectively. In Table 2 are cases when the channel is terminated in high impedances, characteristic impedances, lower impedances and averaged impulse response, respectively. It can be observed that the averaged delay spread for a channel with eight branches is 1.5323 μs; this corresponds with 90% coherence bandwidth of 13.052 kHz.

5.1.3 Channel with 12 distributed branches

Consider the configuration with distributed branches as in Fig. 1 with $Z_s = Z_L = 589\ \Omega$. The number of distributed branches between point A and L is 12. The length A–L was kept constant and equal to 4 km. The branch line length was kept

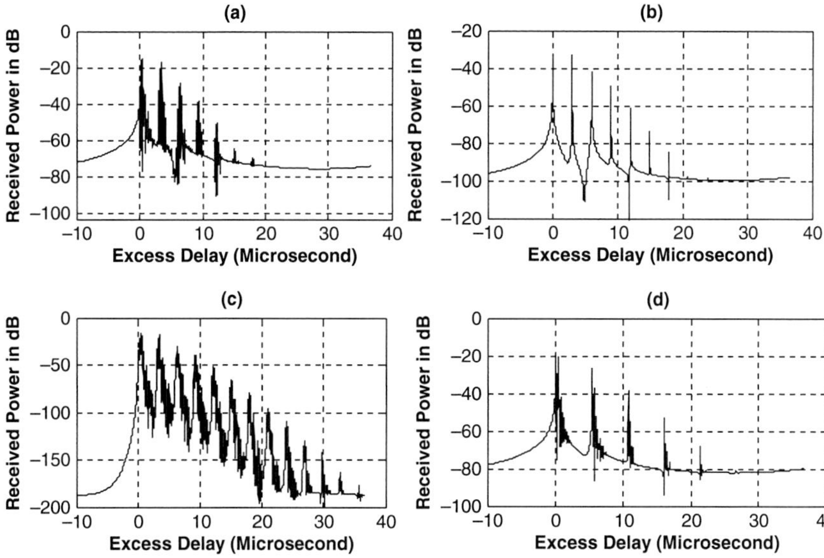

Figure 5: Simulated multipath power delay profiles for medium voltage power-line link with eight distributed branches (a) all terminated in 10 kΩ (b) all terminated in characteristic impedances (c) all terminated in 50 Ω (d) averaged delay profile.

Table 2: Channel parameters for an MV network with eight branches –30 dB, 0.1–100 MHz.

Branch loads	τ (μs)	σ_τ (μs)	$B_{C-90\%}$ (kHz)	$B_{C-50\%}$ (MHz)
10 kΩ	1.7172	1.6312	12.261	0. 12261
50 Ω	1.7050	1.652	12.10	0.1210
Char. Impedance	2.1146	1.875	10.667	0.10667
Average	1.5863	1.5323	13.052	0.13052

at 30 m. The terminal loads at distributed branches were varied as terminated in (50 Ω, characteristic impedance and 10 kΩ). Note that the branches were equally distributed between A and L. Figure 6 shows the multipath channel impulse response of a power-line link with 12 distributed branches. Figure 7 shows the delay profiles of power-line link with 12 branches and terminated in various loads impedances. Figure 7d is averaged delay profile which is obtained after averaging the responses of Fig. 7a–c. It should be noted that the attenuations of multipaths increase as the channel is terminated in characteristic impedances. From Fig. 7, the channel characterization parameters were calculated and are as indicated in Table 3. The threshold of –30 dB from the maximum value was considered; the maximum excess delay τ_m is varying between 10, 10.2, 9 and 10 μs for a channel

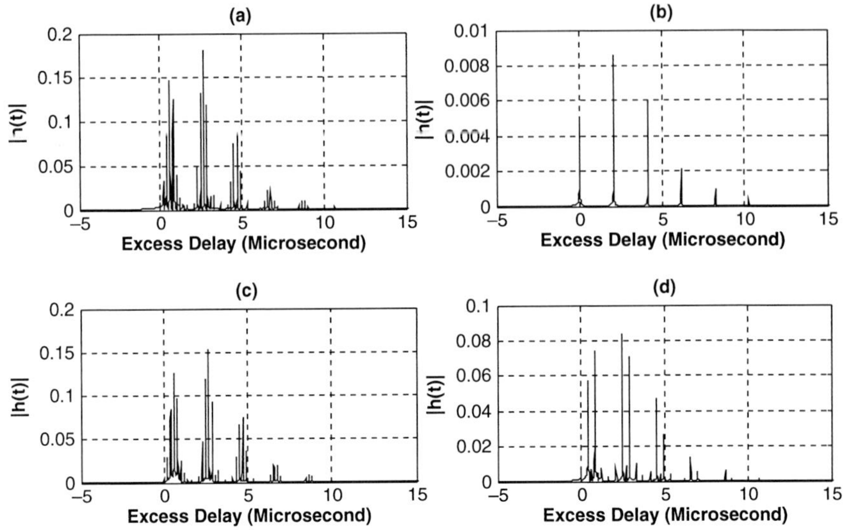

Figure 6: Simulated multipath impulse response profiles for medium voltage power-line link with 12 distributed branches (a) all terminated in 10 kΩ (b) all terminated in characteristic impedances (c) all terminated in 50 Ω (d) averaged delay profile.

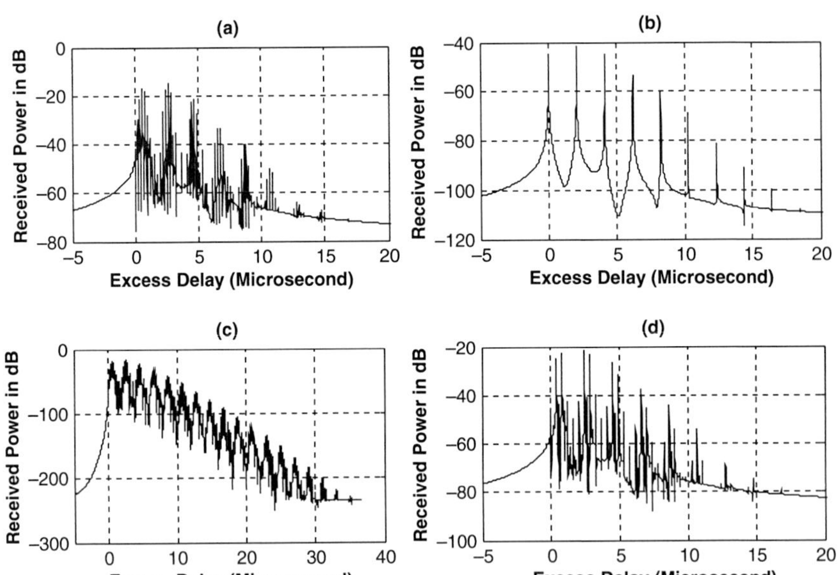

Figure 7: Simulated multipath power delay profiles for medium voltage power-line link with 12 distributed branches (a) all terminated in 10 kΩ (b) all terminated in characteristic impedances (c) all terminated in 50 Ω (d) averaged delay profile.

Table 3: Channel parameters for an MV network with 12 branches –30 dB, 0.1–100 MHz.

Branch loads	τ (μs)	σ_τ (μs)	$B_{C\text{–}90\%}$ (kHz)	$B_{C\text{–}50\%}$ (MHz)
10 kΩ	2.1655	1.4009	14.277	0.14277
50 Ω	2.2307	1.5614	12.809	0.12809
Char. Impedance	2.5332	1.5482	12.918	0.12918
Average	2.177	1.4289	13.996	0.13996

terminated in high impedances, low impedances, characteristic impedances and averaged channel response, respectively. Table 3 shows cases when the channel is terminated in high impedances, lower impedances, characteristic impedances and averaged impulse response, respectively. It can be observed that the averaged delay spread for a channel with 12 branches is 1.4289 μs; this corresponds with 90% coherence bandwidth of 13.996 kHz.

5.2 Low voltage systems

The LV power channel similar to the systems found in Tanzania was considered. The line parameters, hence are L_e = 1.9648 μH/m, C_e = 5.6627 pF/m [19, 20]. The investigations considered the network with 4, 8, 12 and 16 branches.

5.2.1 Channel with four distributed branches

Consider the configuration with distributed branches as in Fig. 1 with $Z_s = Z_L$ = 589 Ω. The number of distributed branches between point A and L is four. The length A–L was kept constant and equal to 1.2 km. The branch line lengths were kept at 20 m. The terminal loads at distributed branches were varied as terminated in (50 Ω, characteristic impedance and 10 kΩ). Note that the branches were equally distributed between A and L.

For each case the important channel parameters were investigated. Figure 8 shows the multipath power delay profiles of a power-line link with four distributed branches. Figure 9 shows the delay profiles of power-line link with four branches and terminated in various loads impedances. Figure 9d is averaged delay profile which is obtained after averaging the responses of Fig. 9a–c. It should be noted that the attenuations of multipaths increase as the channel is terminated in characteristic impedances.

From Fig. 9, the channel characterization parameters were calculated and are as indicated in Table 4. The threshold of –30 dB from the maximum value was considered; the maximum excess delay τ_m is 11 μs for a channel terminated impedances cases. In Table 4 are cases when the channel is terminated in high impedances, characteristic impedances, lower impedances and averaged impulse response, respectively. It can be observed that the averaged delay spread for a channel with

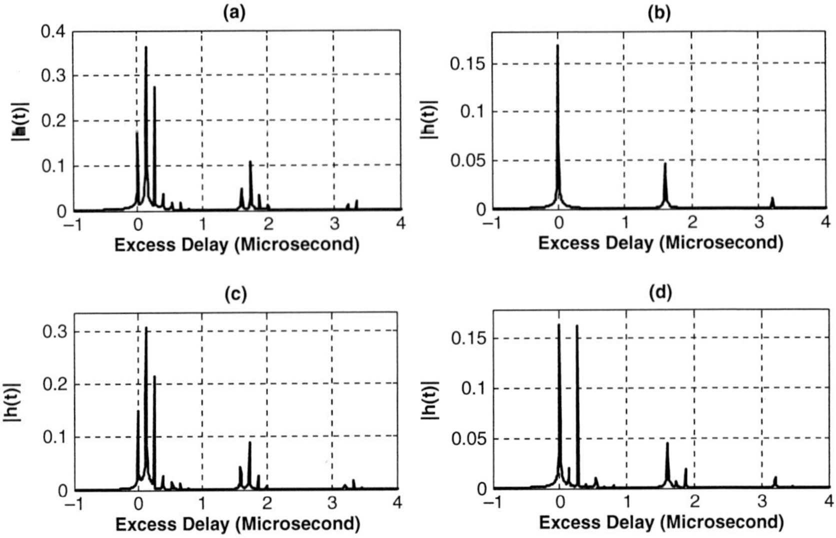

Figure 8: Simulated multipath impulse response for low voltage power-line link with four distributed branches (a) all terminated in 10 kΩ (b) all terminated in characteristic impedances (c) all terminated in 50 Ω (d) averaged response.

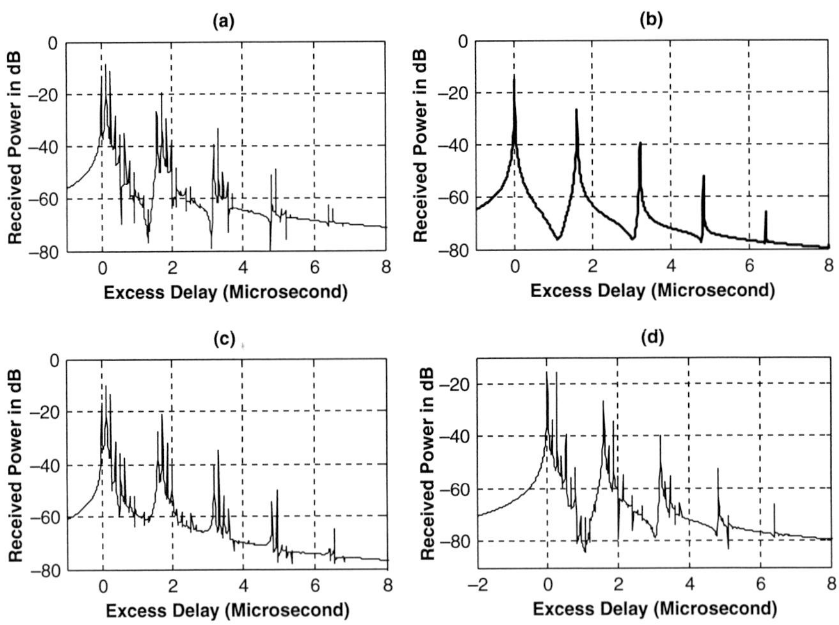

Figure 9: Simulated multipath power delay profiles for low voltage power-line link with four distributed branches (a) all terminated in 10 kΩ (b) all terminated in characteristic impedances (c) all terminated in 50 Ω (d) averaged delay profile.

Table 4: Channel parameters for an LV network with four branches –30 dB, 0.1–100 MHz.

Branch loads	τ (μs)	σ_τ (μs)	$B_{C-90\%}$ (kHz)	$B_{C-50\%}$ (MHz)
10 kΩ	0.2555	0.38288	52.236	0.52236
50 Ω	0.2595	0.40413	49.489	0.49489
Char. Impedance	0.1259	0.45149	44.298	0.44298
Average	0.2060	0.36284	55.120	0.55120

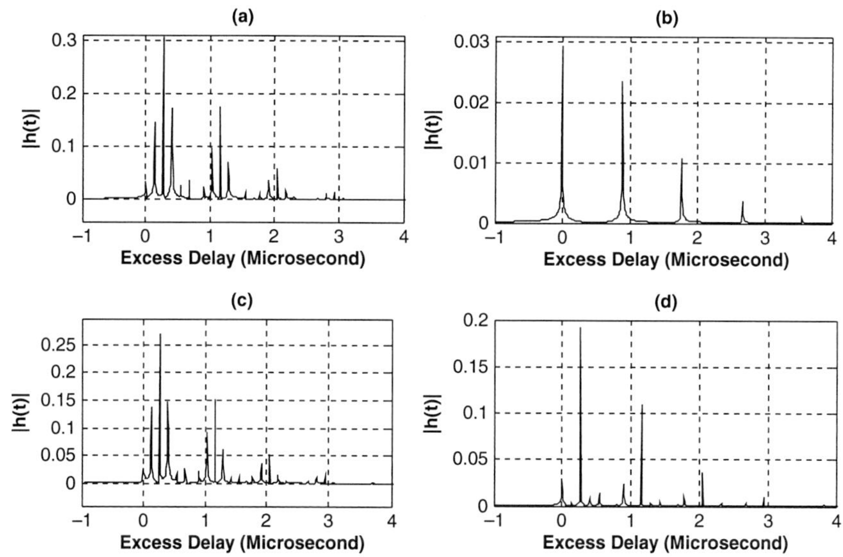

Figure 10: Simulated multipath impulse response profiles for low voltage powerline link with eight distributed branches (a) all terminated in 10 kΩ (b) all terminated in characteristic impedances (c) all terminated in 50 Ω (d) averaged delay profile.

four branches is 0.36284 μs; this corresponds with 90% coherence bandwidth of 55.12 kHz.

5.2.2 Channel with eight distributed branches

Consider a configuration with distributed branches as in Fig. 1 with $Z_s = Z_L = 589\ \Omega$. The number of distributed branches between point A and L is eight. The length A–L was kept constant and equal to 1.2 km. The branch line length was kept at 30 m. The terminal loads at distributed branches were varied as terminated in (50 Ω, characteristic impedance and 10 kΩ). Note that the branches were equally distributed between A and L. For each case the important channel parameters were investigated. Figure 10 shows the multipath power delay profiles of a powerline link with eight distributed branches. Figure 11 shows the delay profiles of power-line link with eight branches and terminated in various loads impedances. Figure 11d is averaged delay profile which is obtained after averaging the responses of Fig. 11a–c. It should be noted that the attenuations of multipaths increase as the channel is terminated in characteristic impedances. From Fig. 11, the channel characterization parameters were calculated and are as indicated in Table 5. The threshold of –30 dB from the maximum value was considered; the maximum excess delay τ_m is varying between 10, 11.9, 10 and 9.3 μs for a channel terminated in high impedances, characteristic impedances, low impedances and averaged channel response, respectively. In Table 5 are cases when the channel is terminated in high impedances, characteristic impedances, lower impedances

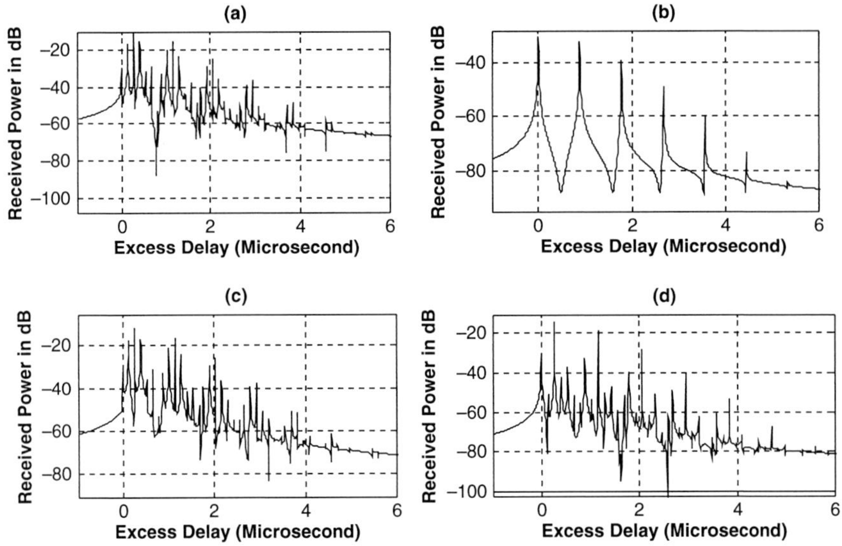

Figure 11: Simulated multipath power delay profiles for low voltage power line link with eight distributed branches (a) all terminated in 10 kΩ (b) all terminated in characteristic impedances (c) all terminated in 50 Ω (d) averaged delay profile.

Table 5: Channel parameters for an LV network with eight branches –30 dB, 0.1–100MHz.

Branch loads	τ (μs)	σ_τ (μs)	$B_{C-90\%}$ (kHz)	$B_{C-50\%}$ (MHz)
10 kΩ	0.529	0.46249	43.2	0.43244
50 Ω	0.52	0.45566	43.892	0.43892
Char. Impedance	0.5951	0.35485	56.362	0.56362
Average	0.5338	0.4694	42.606	0.42606

and averaged impulse response, respectively. It can be observed that the averaged delay spread for a channel with eight branches is 0.4694 μs; this corresponds with 90% coherence bandwidth of 42.606 kHz.

5.2.3 Channel with 12 distributed branches

Consider a configuration with distributed branches as in Fig. 1 with $Z_s = Z_L = 589\ \Omega$. The number of distributed branches between point A and L is 12. The length A–L was kept constant and equal to 1.2 km. The branch line length was kept at 30 m. The terminal loads at distributed branches were varied as terminated in (50 Ω, characteristic impedance and 10 kΩ). Note that the branches were equally distributed between A and L. For each case the important channel parameters were investigated. Figure 12 shows the multipath power delay profiles of a power-line link with 12 distributed branches. Figure 13 shows the delay profiles of power-line link with 12 branches and terminated in various loads impedances. Figure 13d is averaged delay profile which is obtained after averaging the responses of Fig. 13a–c. It should be noted that the attenuations of multipaths increase as the channel is terminated in characteristic impedances. From Fig. 13, the channel characterization parameters were calculated and are as indicated in Table 6. The threshold of –30 dB from the maximum value was considered; the maximum excess delay τ_m is varying between 10, 10.2, 9 and 10 μs for a channel terminated in high impedances, low impedances, characteristic

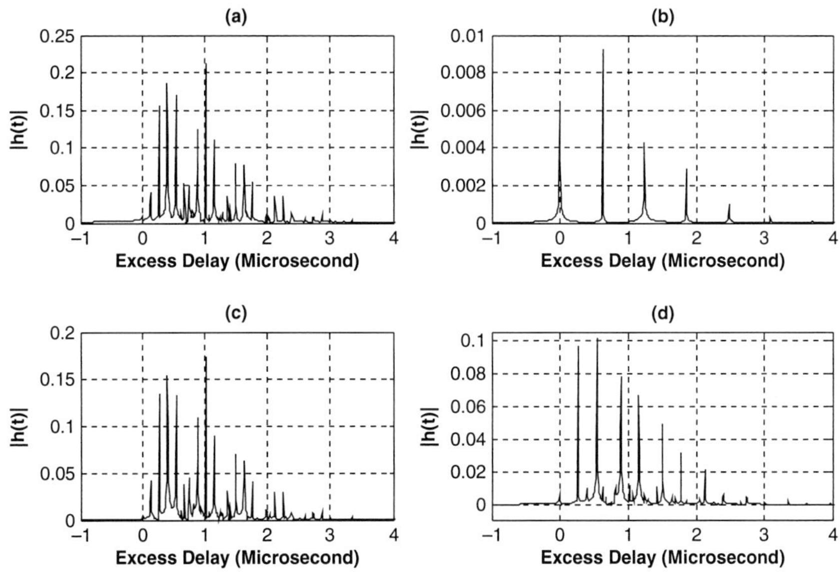

Figure 12: Simulated multipath impulse response profiles for low voltage power-line link with 12 distributed branches (a) all terminated in 10 kΩ (b) all terminated in characteristic impedances (c) all terminated in 50 Ω (d) averaged delay profile.

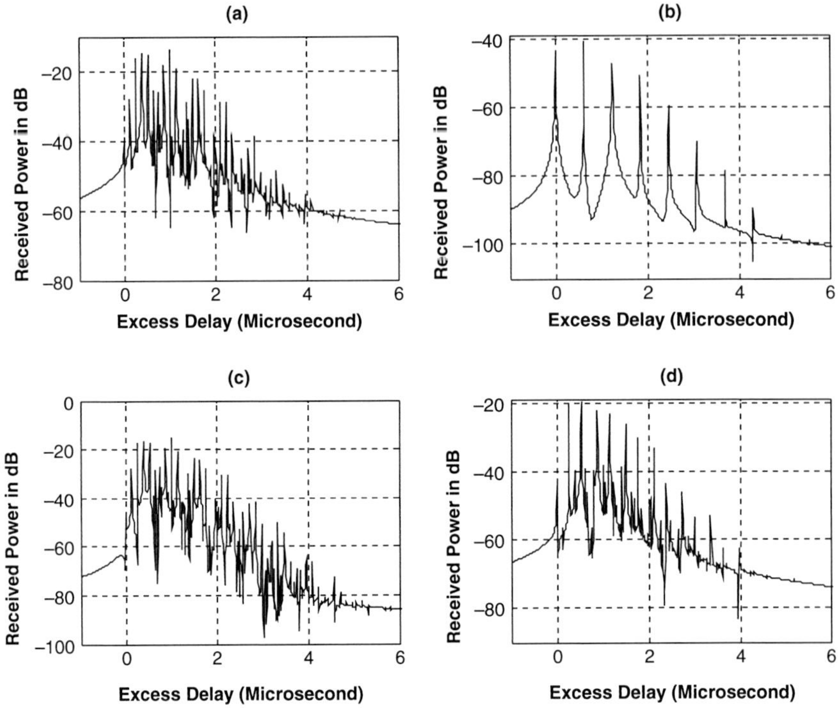

Figure 13: Simulated multipath power delay profiles for low voltage power-line link with 12 distributed branches (a) all terminated in 10 kΩ (b) all terminated in characteristic impedances (c) all terminated in 50 Ω (d) averaged delay profile.

Table 6: Channel parameters for an LV network with 12 branches –30 dB, 0.1–100 MHz.

Branch loads	τ (µs)	σ_τ (µs)	$B_{C\text{–}90\%}$ (kHz)	$B_{C\text{–}50\%}$ (MHz)
10 kΩ	0.7879	0.43707	45.759	0.45759
50 Ω	0.7774	0.43507	45.969	0.45969
Char. Impedance	0.6299	0.50466	39.631	0.39631
Average	0.5656	0.49333	40.541	0.40541

impedances and averaged channel response, respectively. In Table 6 are cases when the channel is terminated in high impedances, lower impedances, characteristic impedances and averaged impulse response, respectively. It can be observed that the averaged delay spread for a channel with 12 branches is 0.49333 µs; this corresponds with 90% coherence bandwidth of 40.541 kHz.

5.3 Indoor systems power-line channel analysis

The IV power-line channel similar to the systems in Tanzanian residences or offices was considered. The line parameters are L_e = 0.44388 μH/m, C_e = 61.734 pF/m [19, 20]. The investigations considered the network with 4, 8 and 12 branches.

5.3.1 Channel with four distributed branches

Consider a configuration with distributed branches as in Fig. 1 with $Z_s = Z_L = 85\ \Omega$. The number of distributed branches between point A and L is four. The length A–L was kept constant and equal to 20 m. The branch line lengths were kept at 10 m. The terminal loads at distributed branches were varied as terminated in (50 Ω, characteristic impedance and 500 Ω). Note that the branches were equally distributed between A and L. Figure 14 shows the multipath power delay profiles of a power-line link with four distributed branches.

Figure 15 shows the delay profiles of power-line link with four branches and terminated in various loads impedances. Figure 15d is averaged delay profile which is obtained after averaging the responses of Fig. 15a–c. It should be noted that the attenuations of multipaths increase as the channel is terminated in characteristic impedances. From Fig. 15, the channel characterization parameters were calculated and are as indicated in Table 7. The threshold of –30 dB from the maximum value was considered; the maximum excess delay τ_m is varying between 0.41, 0.25, 0.20 and 0.41 μs for a channel terminated in high impedances, low impedances, characteristic impedances and averaged channel response, respectively. In Table 7 are

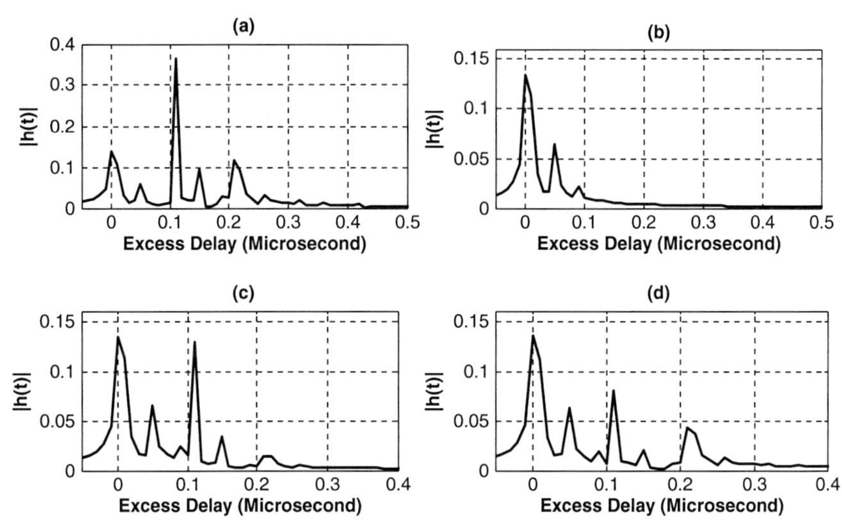

Figure 14: Simulated multipath impulse response for indoor power-line link with four distributed branches (a) all terminated in 500 Ω (b) all terminated in characteristic impedances (c) all terminated in 50 Ω (d) averaged response.

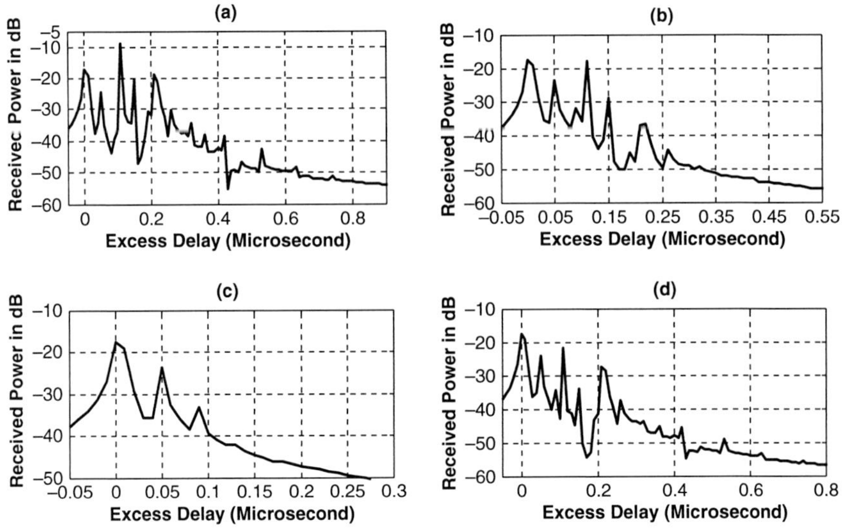

Figure 15: Simulated multipath power delay profiles for indoor power-line link with four distributed branches (a) all terminated in 500 Ω (b) all terminated in 50 Ω (c) all terminated in characteristic impedances (d) averaged delay profile.

Table 7: Channel parameters for an IV network with four branches –30 dB, 0.1–100 MHz.

Branch loads	τ (μs)	σ_τ (μs)	$B_{C\text{–}90\%}$ (kHz)	$B_{C\text{–}50\%}$ (MHz)
10 kΩ	0.1245	0.0879	227.67	2.2767
50 Ω	0.0739	0.0641	312.22	3.1222
Char. Impedance	0.0236	0.03132	638.48	6.3848
Average	0.0902	0.09549	209.45	2.0945

cases when the channel is terminated in high impedances, lower impedances, characteristic impedances and averaged impulse response, respectively. It can be observed that the averaged delay spread for a channel with four branches is 0.09549 μs; this corresponds with 90% coherence bandwidth of 209.45 kHz.

5.3.2 Channel with eight distributed branches

Consider a configuration with distributed branches as in Fig. 1 with $Z_s = Z_L = 85\ \Omega$. The number of distributed branches between point A and L is eight. The length A–L was kept constant and equal to 20 m. The branch line length was kept at 10 m. The terminal loads at distributed branches were varied as terminated in (50 Ω, characteristic impedance and 500 Ω). Note that the branches were equally distributed between A and L. For each case the important channel parameters were investigated. Figure 16 shows the multipath power delay profiles of a power-line link with eight distributed branches. Figure 17 shows the delay profiles of power-line link with eight branches and terminated in various load impedances.

Figure 17d is averaged delay profile which is obtained after averaging the responses of Fig. 17a–c. It should be noted that the attenuations of multipaths increase as the channel is terminated in characteristic impedances. From Fig. 17, the channel characterization parameters were calculated and are as indicated in Table 8. The threshold of –30 dB from the maximum value was considered; the maximum excess delay τ_m is varying between 0.6, 0.3, 0.15 and 0.5 μs for a channel terminated in high impedances, low impedances, characteristic impedances and averaged channel response, respectively. In Table 8 are cases when the channel is terminated in high impedances, lower impedances, characteristic impedances and averaged impulse response, respectively. It can be observed that the averaged delay spread for a channel with eight branches is 0.118 μs; this corresponds with 90% coherence bandwidth of 170.22 kHz.

5.3.3 Channel with 12 distributed branches

Consider a configuration with distributed branches as in Fig. 1 with $Z_s = Z_L = 85\ \Omega$. The number of distributed branches between point A and L is 12. The length A–L was kept constant and equal to 20 m. The branch line length was varied kept at 10 m. The terminal loads at distributed branches were varied as terminated in (50 Ω,

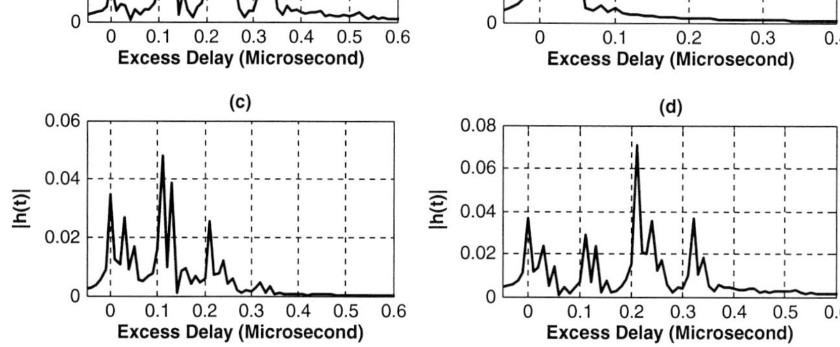

Figure 16: Simulated multipath impulse response profiles for indoor power-line link with eight distributed branches (a) all terminated in 500 Ω (b) all terminated in characteristic impedances (c) all terminated in 50 Ω (d) averaged delay profile.

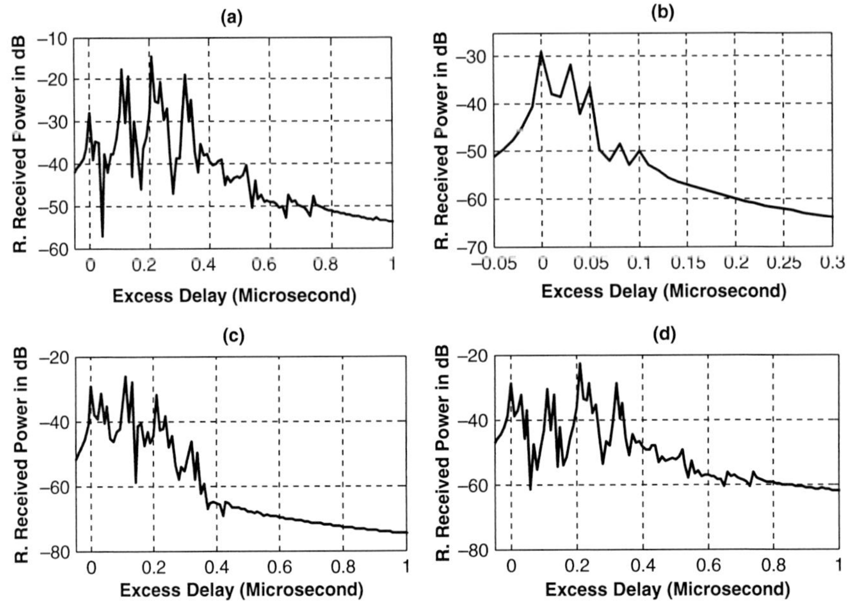

Figure 17: Simulated multipath power delay profiles for indoor power-line link with eight distributed branches (a) all terminated in 500 Ω (b) all terminated in characteristic impedances (c) all terminated in 50 Ω (d) averaged delay profile.

Table 8: Channel parameters for an IV network with eight branches –30 dB, 0.1–100 MHz.

Branch loads	τ (μs)	σ_τ (μs)	$B_{C-90\%}$ (kHz)	$B_{C-50\%}$ (MHz)
10 kΩ	0.207	0.107	187.15	1.8715
50 Ω	0.1174	0.0866	230.85	2.3085
Char. Impedance	0.0206	0.0273	731.56	7.3156
Average	0.183	0.118	170.22	1.7022

characteristic impedance and 500 Ω). Note that the branches were equally distributed between A and L. Figure 18 shows the multipath power delay profiles of a power-line link with 12 distributed branches. Figure 19 shows the delay profiles of power-line link with 12 branches and terminated in various loads impedances. Figure 19d is averaged delay profile which is obtained after averaging the responses of Fig. 19a–c.

It should be noted that the attenuations of multipaths increase as the channel is terminated in characteristic impedances. From Fig. 19, the channel characterization parameters were calculated and are as indicated in Table 9. The threshold of –30 dB from the maximum value was considered; the maximum excess delay τ_m is varying between 0.75, 0.45, 0.15 and 0.75 μs for a channel terminated in high impedances,

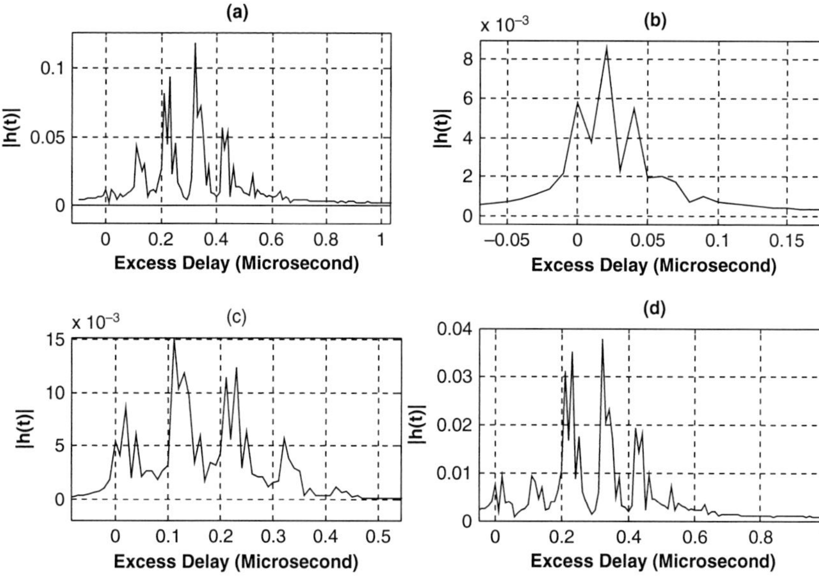

Figure 18: Simulated multipath impulse response profiles for indoor power-line link with 12 distributed branches (a) all terminated in 500 Ω (b) all terminated in characteristic impedances (c) all terminated in 50 Ω (d) averaged delay profile.

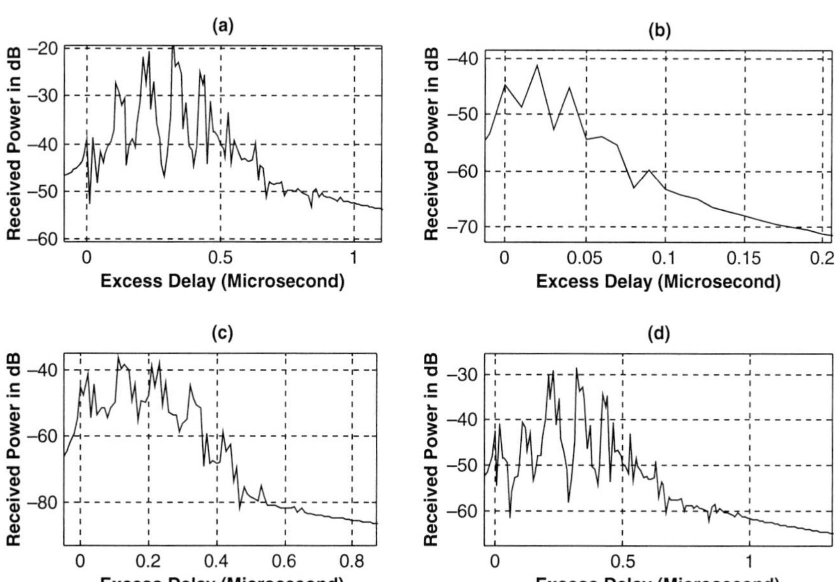

Figure 19: Simulated multipath power delay profiles for indoor power-line link with 12 distributed branches (a) all terminated in 500 Ω (b) all terminated in characteristic impedances (c) all terminated in 50 Ω (d) averaged delay profile.

low impedances, characteristic impedances and averaged channel response, respectively. In Table 9 are cases when the channel is terminated in high impedances, lower impedances, characteristic impedances and averaged impulse response, respectively. It can be observed that the averaged delay spread for a channel with four branches is 0.132 μs; this corresponds with 90% coherence bandwidth of 151.21 kHz.

Table 9: Channel parameters for an IV network with 12 branches –30 dB, 0.1–100 MHz.

Branch loads	τ (μs)	σ_τ (μs)	$B_{C\text{–}90\%}$ (kHz)	$B_{C\text{–}50\%}$ (MHz)
10 kΩ	0.2967	0.128	156.12	1.5612
50 Ω	0.1526	0.0968	206.56	2.0656
Char. Impedance	0.0263	0.0226	884.48	8.8448
Average	0.2922	0.132	151.21	1.5121

5.3.4 Channel with 16 distributed branches

Consider a configuration with distributed branches as in Fig. 1 with $Z_s = Z_L = 85\ \Omega$. The number of distributed branches between point A and L is 16. The length A–L was kept constant and equal to 20 m. The branch line length was kept at 10 m. The terminal loads at distributed branches were varied as terminated in (50 Ω, characteristic impedance and 500 Ω). Note that the branches were equally distributed between A and L. For each case the important channel parameters were investigated. Figure 20

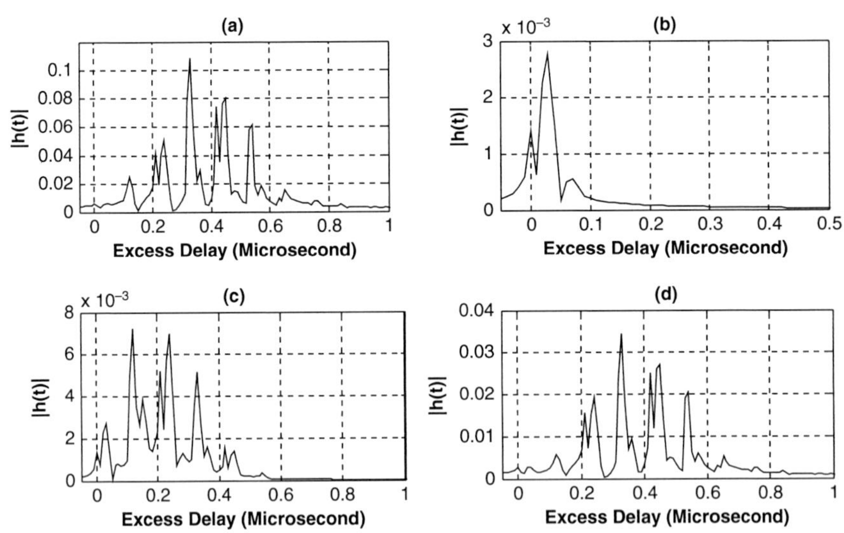

Figure 20: Simulated multipath impluse response profiles for indoor power-line link with 16 distributed branches (a) all terminated in 500 Ω (b) all terminated in characteristic impedances (c) all terminated in 50 Ω (d) averaged delay profile.

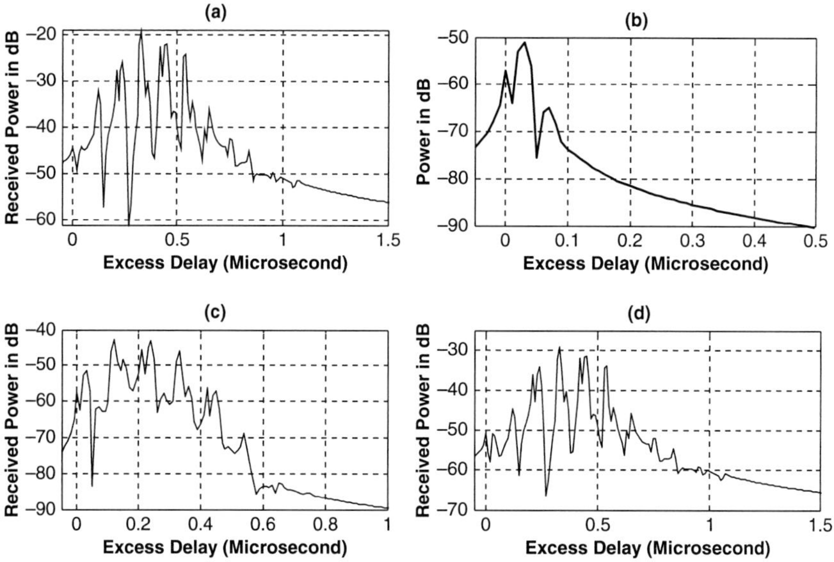

Figure 21: Simulated multipath power delay profiles for indoor power-line link with 16 distributed branches (a) all terminated in 500 Ω (b) all terminated in characteristic impedances (c) all terminated in 50 Ω (d) averaged delay profile. (© [2009] IEEE)

Table 10: Channel parameters for an IV network with 16 branches –30 dB, 0.1–100 MHz.

Branch loads	τ (μs)	σ_τ (μs)	$B_{C\text{–}90\%}$ (kHz)	$B_{C\text{–}50\%}$ (MHz)
10 kΩ	0.3984	0.16272	122.91	1.2291
50 Ω	0.2391	0.14208	140.77	1.4077
Char. Impedance	0.0262	0.02123	942.21	9.4221
Average	0.3783	0.14555	137.41	1.3741

shows the multipath power delay profiles of a power-line link with 16 distributed branches. Figure 21 shows the delay profiles of power-line link with 16 branches and terminated in various loads impedances. Figure 21d is averaged delay profile which is obtained after averaging the responses of Fig. 21a–c. It should be noted that the attenuations of multipaths increase as the channel is terminated in characteristic impedances. From Fig. 21, the channel characterization parameters were calculated and are as indicated in Table 10. The threshold of –30 dB from the maximum value was considered; the maximum excess delay τ_m is varying between 1.0, 0.5, 0.20 and 1.0 μs for a channel terminated in high impedances, low impedances, characteristic impedances and averaged channel response, respectively. It can be observed that the averaged delay spread for a channel with four branches is 0.14555 μs; this corresponds with 90% coherence bandwidth of 137.41 kHz.

6 Noise in power-line networks

It has been said that without electromagnetic shielding, power-line cables are sensitive to external noises from various radio frequency devices and electromechanical equipments leading to electromagnetic interference problems. There has been considerable effort by various researchers to model the noise in PLC network. The noise in PLC systems can be classified into three types namely, colored background noise, narrow band noise and impulse noise. Colored noise is the sum total of various noise sources with low power. Narrow band (ingress) noise has amplitude modulated signals caused by induction from radio station signals in medium and short waves bands.

The impulse noise is caused by transients due to switching or lightning phenomena within the power network. The power spectral density (PSD) for background noise is usually around –145 dBm/Hz and this is about 30 dB above thermal noise floor [22]. The impulsive noise has maximum amplitude of 40 dBm/Hz higher than background and/or narrow band noise. These conclusions are made based on a noise model developed by Chen [22], which was obtained based on measurements of a typical indoor channel by Liu *et al.* [23]. This typical noise signature in power-line network can be represented as shown in Fig. 22. Since our investigation was to determine the sensitivity factors which influence power-line channel capacity based on network configuration/infrastructure, the same noise model was used throughout even for low voltage and medium channels. Next let us use the channel models and noise in power line to determine the channel capacities in different

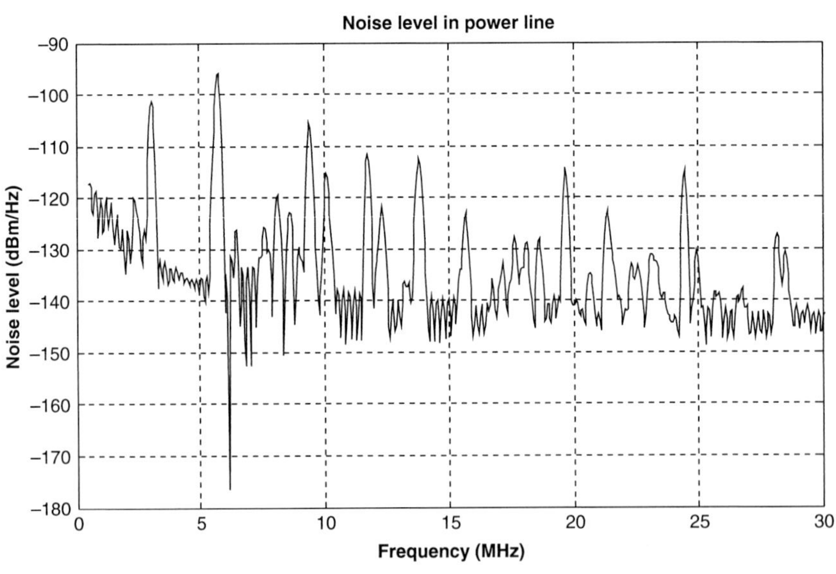

Figure 22: Noise in power-line network (© [2009] IEEE) [21–22].

power-line channels (medium voltage, low voltage and indoor channel) for various cases of branches, lengths and terminal load conditions.

7 Channel capacities for different PLC links

Consider the distributed branches network as shown in Fig. 23. The number of branches was varied in the link between points A and J. The distance between points A and J was 20 m, 4 km and 1.2 km, while all corresponding branch lengths were 10, 30 and 20 m long for IV channel, MV channel and LV channel, respectively. The distribution of number of branches was varied as 2, 5, 10 and 15 for MV and LV channels, while for the IV channel it was varied as 4, 8, 12 and 16. The per unit length parameters for medium voltage line, low voltage line and indoor power-line are L_e = 1.9648 μH/m, C_e = 5.6627 pF/m; L_e = 1.9589 μH/m, C_e = 5.6799 pF/m; and L_e = 0.44388 μH/m, C_e = 61.734 pF/m, respectively. Note that for each case the distances between the branches were equal and equally distributed between the link A and J. Now for each branch case corresponding to the branch length and number of branches the load impedance was varied as 5 Ω, 50 Ω, characteristic impedance, 500 Ω and 5 kΩ. For any case treated the channel transfer function *H*(*f*) is calculated as discusses in Chapter 3.

The channel capacity was determined using (13) [19], whereby the frequency variation was between 1 and 30 MHz. *S*(*f*) is the received signal power and *N*(*f*) is the noise power, which is dependent on transmitted signal power and channel transfer function as given in (14). Noise power level *N*(*f*) for different frequencies is considered based on Fig. 22. Due to the limitations on the transmitted power, the field strength is limited to 30 dBμV/m. Thus allowed PSD can be estimated according to (15) and is found to be between –72 and –52 dBm/Hz corresponding to a coupling factor [24] in the range of –65 and –45 dB for a distance of about 30 m. For the study we chose the range of PSD to be between –90 and –30 dBm/Hz.

$$C = \int_{f1}^{f2} \log_2 \left[1 + \frac{S(f)}{N(f)} \right] df \qquad (13)$$

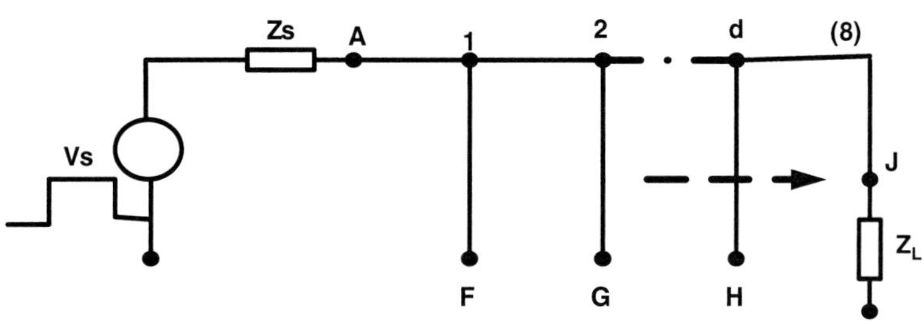

Figure 23: Power-line medium voltage network with distributed branches.

$$S(f) = PSD \cdot |H(f)|^2 \tag{14}$$

$$PSD = (-CF - 117)(dBm / Hz) \tag{15}$$

Figure 24 shows the variation of channel capacity in Mbps against PSD in dBm/Hz for IV channels for various branch numbers and different load terminations. It can be observed from Fig. 24 that channel capacity decreases with increase in number of branches between the sending and receiving ends. For e.g. at a PSD of –60 dBm/Hz (which is a typical PSD PLC device in the market) [15] the channel capacity of the indoor channel with 4 and 16 branches terminated in characteristic impedance are 600 and 300 Mbps, respectively. For the link with less number of branches the influence of load impedance is negligible. However, it is seen that as the number of branches increases more than eight, the influence of branches on channel capacity is predominant. It is interesting to observe that the channel capacity is minimum when the load impedances are terminated in characteristic impedance. The differences in the channel capacity are negligible whether the load is 5 Ω or 5 kΩ. It seems that the terminal load variations do not affect the channel capacity for indoor channels excepting with characteristic impedance terminations. But the channel capacity is lesser with larger branches. To demonstrate

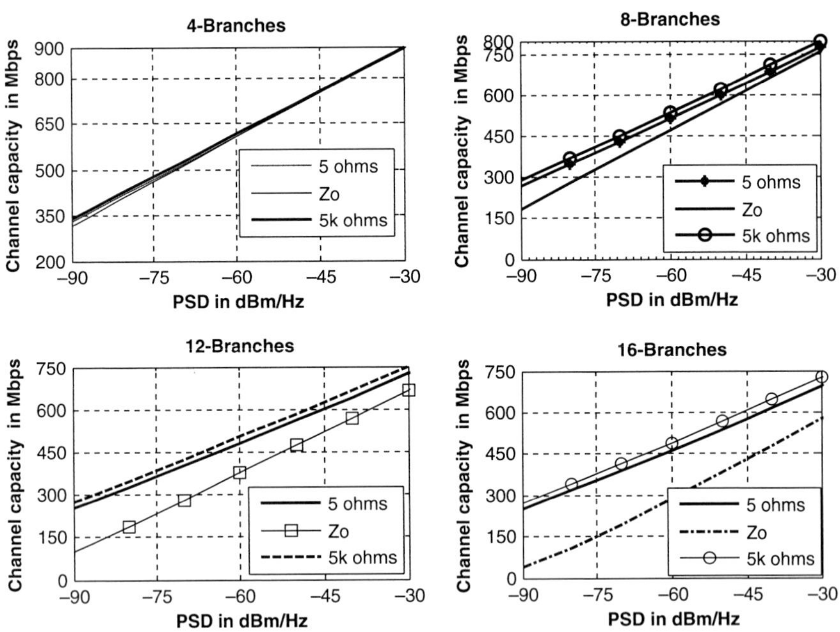

Figure 24: The channel capacity for indoor power-line network with (a) 4 (b) 8 (c) 12 (d) 16 branches, with different branch terminations (© [2009] IEEE).

the sensitivity of channel capacity variations with terminal load variations and number of branches corresponding to Fig. 24, it is seen for e.g. at –60 dBm/Hz, the difference in channel capacity for 4, 8, 12 and 16 branches are 0, 50, 100 and 200Mbps, respectively. This concludes that we lose about 200 Mbps if the number of branches were increased by four times.

Figure 25 shows the variation of channel capacity against PSD for different branch numbers and with different branch lengths. Here we keep the terminal loads at 50 Ω and the length of the link is 20 m. The branch lengths are varied as 5, 10, 15 and 20 m. Similar to the earlier case as the number of branches increase the channel capacity decreases. For e.g. an increase in number of branches by three times can lead to a loss of 250 Mbps in channel capacity for any of the case that is treated in Fig. 25. This concludes that the influence of branch length on channel capacity is negligible.

Figure 26 shows channel capacity against PSD for different number of branches and at different load impedances for a medium voltage channel (link was 4 km) and the branch lengths were 30 m. The observations are very similar to the corresponding case of indoor channel. Figure 27 shows the variation of channel capacity against PSD for different number of branches and with different branch lengths for medium voltage channel. The termination for this case was fixed at 5 Ω. Again the observations are similar to the corresponding case of the indoor channel.

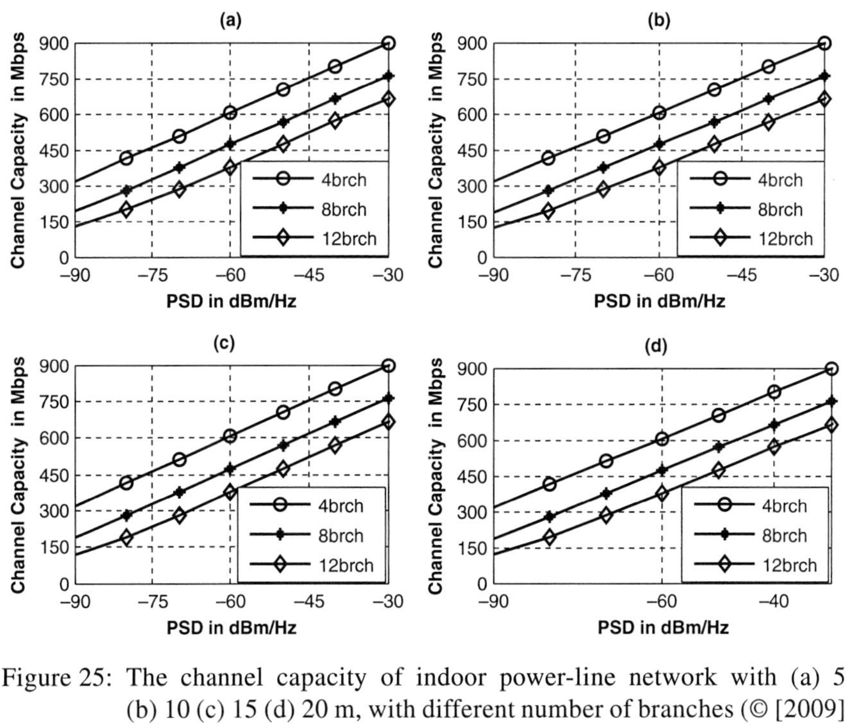

Figure 25: The channel capacity of indoor power-line network with (a) 5 (b) 10 (c) 15 (d) 20 m, with different number of branches (© [2009] IEEE).

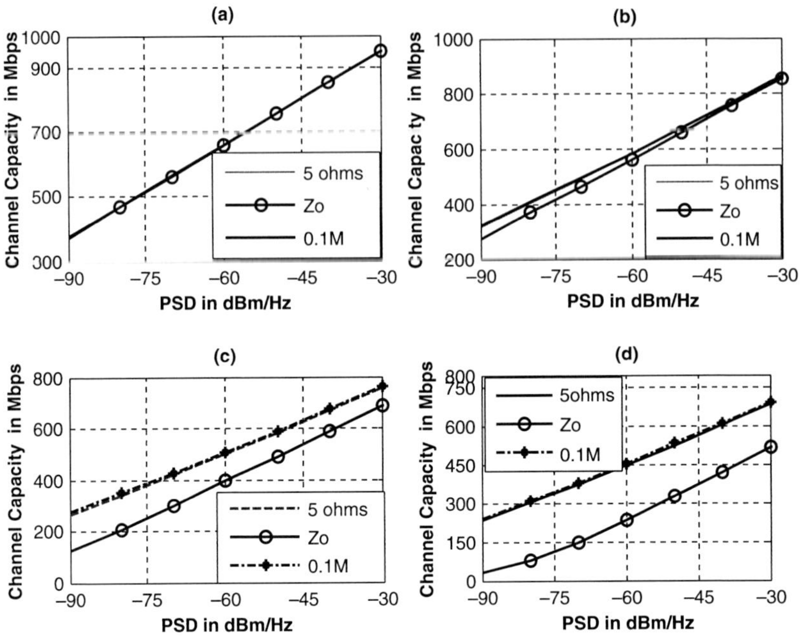

Figure 26: The channel capacity of medium voltage channel (a) 2 (b) 5 (c) 10 (d) 15 branches, with different branch load terminations (© [2009] IEEE).

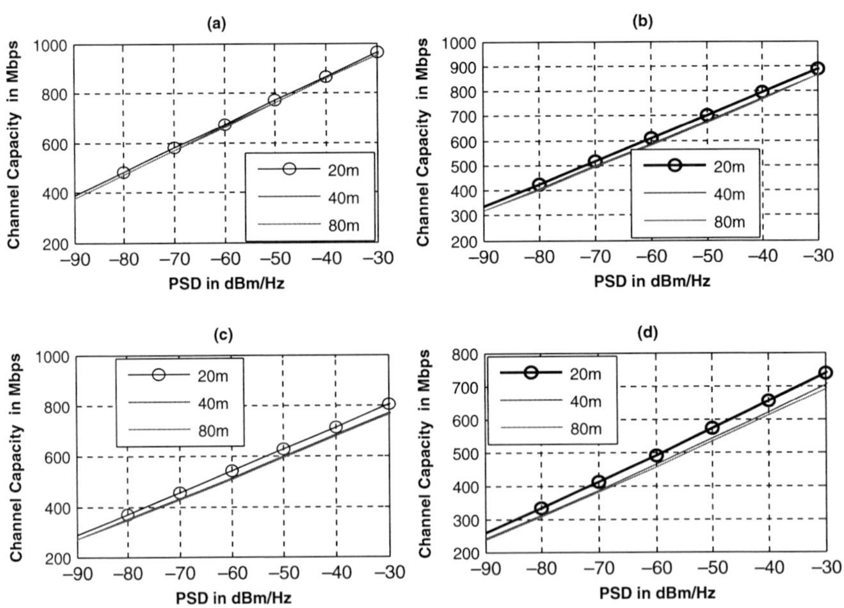

Figure 27: The channel capacity of medium voltage channel (a) 2 (b) 5 (c) 10 (d) 15 branches, with different branch lengths (© [2009] IEEE).

Figure 28 shows channel capacity against PSD for different number of branches and at different load impedances for a low voltage channel. The length of the branch length was kept constant at 20 m. Again the observations are similar to corresponding cases of either the MV channel or IV channel. Figure 29 shows the

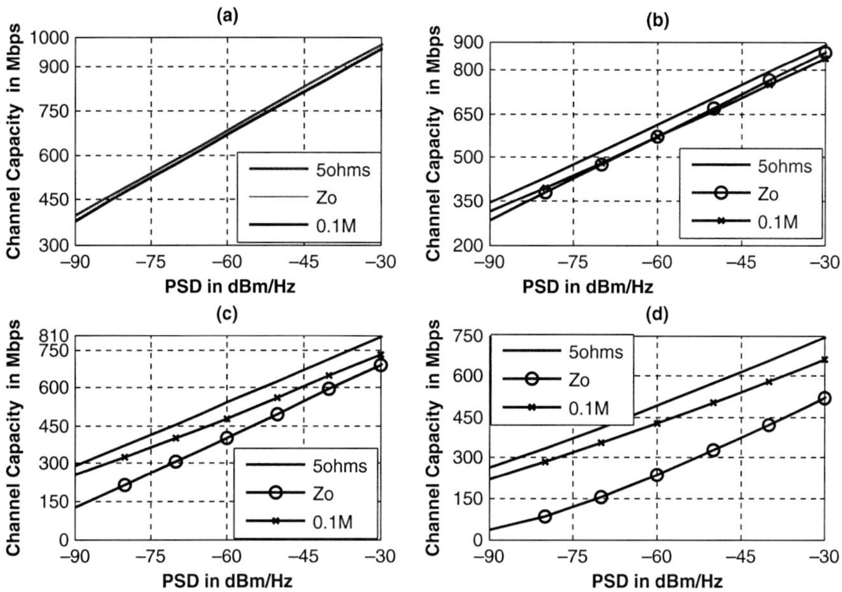

Figure 28: The channel capacity of low voltage channel (a) 2 (b) 5 (c) 10 (d) 15 branches, with different branch loads (© [2009] IEEE).

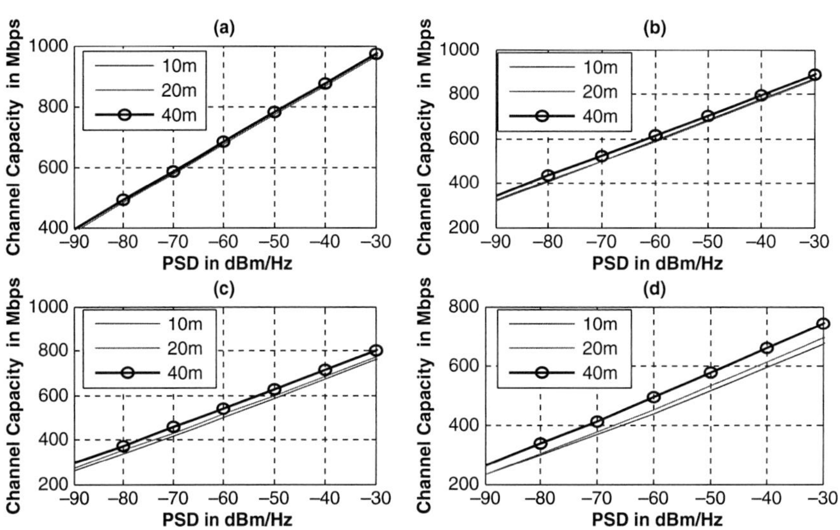

Figure 29: The channel capacity of low voltage channel (a) 2 (b) 5 (c) 10 (d) 15 branches, with different branch lengths (© [2009] IEEE).

variation of channel capacity against PSD for different number of branches and with different branch lengths for low voltage channel. The termination for this case was fixed at 5 Ω. Again the observations are similar to corresponding cases of either the MV channel or IV channel. Next let us consider the medium voltage channel with ground return.

8 The influence of ground return on channel capacity for medium voltage channel

This study is included here in order to predict the feasibilities of the BPLC equipment with ground returns [24]. In this investigation a typical MV channel corresponding to the Tanzanian power-line network is considered. The length of the line was chosen as 4 km, and the number of branches and their layout is similar to the case treated in the previous section for the medium voltage channel. The branch length was kept constant at 30 m. We consider the ground conductivity of either 10 or 1 mS/m as the case may be. The ground relative permittivity was chosen as 10, which is a typical value for most of the ground conditions.

Figure 30 shows the variation of channel capacity in Mbps against PSD in dBm/Hz for various branch numbers and different load terminations for ground conductivity of 10 mS/m. It can be observed from Fig. 30, that channel capacity decreases with increase in number of branches between the sending and receiving ends.

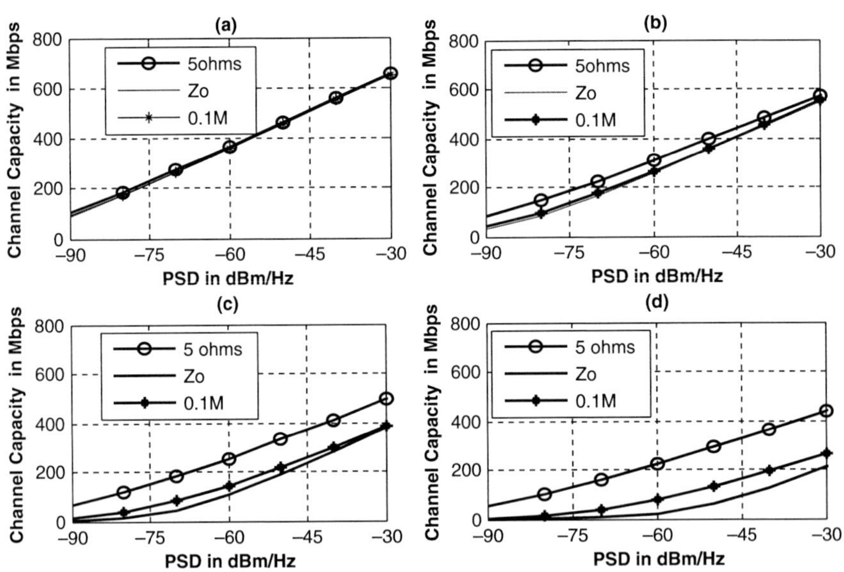

Figure 30: The channel capacity of medium voltage channel (a) 2 (b) 5 (c) 10 (d) 15 branches for 10 mS/m, with different load terminations (© [2009] IEEE).

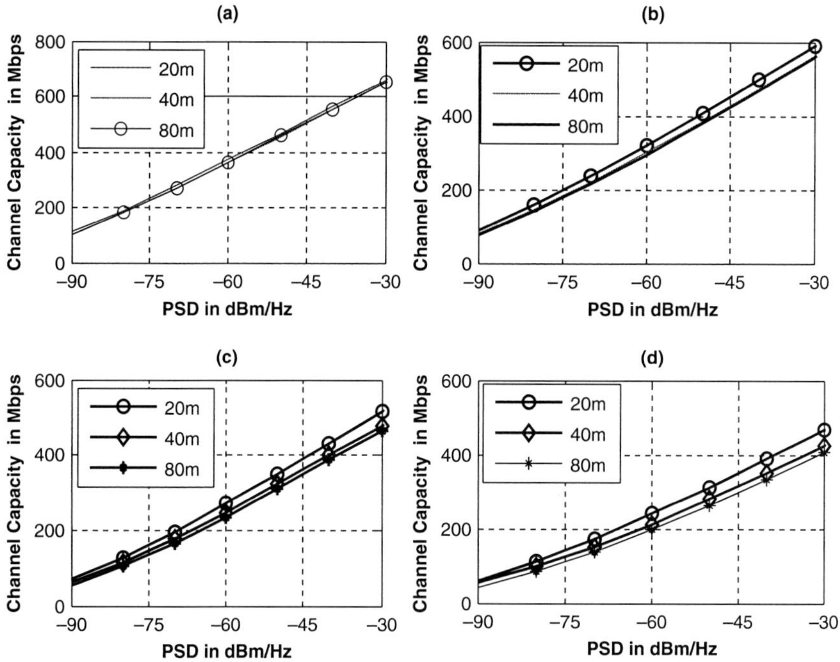

Figure 31: The channel capacity of medium voltage channel (a) 2 (b) 5 (c) 10 (d) 15 branches for 10 mS/m, with different branch lengths (© [2009] IEEE).

For the link with less number of branches the influence of load impedance is negligible as observed in the previous section corresponding to the MV channel with adjacent conductor return. However, it is seen that as the number of branches increases more than five, the influence of branches on channel capacity is predominant and also depends upon the load connected.

If we compare the channel capacity for a network with adjacent conductor return and corresponding ground return case there it is seen that one could lose up to 250 Mbps with ground return. Figure 31 shows the variation of channel capacity against PSD for various numbers of branches and with different branch lengths for ground conductivity of 10 mS/m. We keep the terminal load constant at 5 Ω. The branch lengths are 20, 40 and 80 m as the case may be. Similar to the earlier case as the number of branches increases the channel capacity decreases. For e.g. an increase in number of branches by three times can lead to a loss of 200 Mbps in channel capacity for any of the case that is treated in Fig. 31. However, the influence of branch length on channel capacity seems negligible. Investigations were made with 1 mS/m ground conductivity too. In comparison to Fig. 31, it was observed that the channel capacity was about 200 Mbps lesser compared to 10 mS/m. This indicates poorer the ground conductivity channel capacity decreases.

References

[1] Matthias, G., Rapp, M. & Dostert, K., Power-line channel characteristics and their effect on communication system design. *IEEE Communications Magazine,* pp. 78–86, April 2004.

[2] Esmailian, T., Kschischang, F.R. & Gulak, P.G., In-building power lines as high-speed communication channels: channel characterization and a test channel ensemble. *International Journal on Communication Systems*, pp. 381–400, May 2003.

[3] Prasad, T.V., Srikanth, S., Krishnan, C.N. & Ramakrishna, P.V., Wideband characterization of low voltage outdoor powerline communication channels in India. *Int. Symp. Power-Line Commun. Appl.*, April 2001.

[4] Degardin, V., Lienard, M., Zeddam, A., Gauthier, F. & Degauque, P., Classification and characterization of impulsive noise on indoor power lines used for data communications. *IEEE Trans. Consum. Electron*, 48(4), pp. 913–918, November 2002.

[5] Zimmermann, M. & Dostert, K., A multipath model for the powerline channel. *IEEE Transactions on Communications*, 50(4), pp. 553–559, April 2002.

[6] Homeplug Powerline Alliance, HomePlug AV Specification, Version 1.0.05, October 2006.

[7] Gavette, S., Sharp Labs, HomePlugAV – Detailed Architecture, Homeplug Executive Seminar, November 2005.

[8] Tlich, M., Avril G. & Zeddam, A., Coherence bandwidth and its relationship with the rms delay spread for plc channels using measurements up to 100 MHz. *Proc. IFIP Intl Federation Inf. Processing*, 256, pp. 129–142, December 2007.

[9] Tlich, M., Zeddam, A., Moulin, F. & Gauthier, F., Indoor power-line communications channel characterization up to 100 MHz – part II: time-frequency analysis. *IEEE Transactions on Power Delivery*, 2009.

[10] Liu, E., Gao, Y., Samdani, G., Mukhtar, O. & Korhonen, T., Broadband powerline channel and capacity analysis. *Proc. ISPLC*, pp. 7–11, April 2005.

[11] Anatory, J., Kissaka, M.M. & Mvungi, N.H., Channel model for broadband powerline communication. *IEEE Transactions on Power Delivery*, 22(1), pp. 135–141, January 2007.

[12] Anatory, J. & Theethayi, N., On the efficacy of using ground return in the broadband power line communications – a transmission line analysis. *IEEE Transactions on Power Delivery*, 23(1), pp. 132–139, January 2008.

[13] Anatory, J., Theethayi, N., Thottappillil, R., Kissaka, M.M. & Mvungi, N.H., An experimental validation for broadband power-line communication (BPLC) model. *IEEE Transactions on Power Delivery*, 23(3), pp. 1380–1383, July 2008.

[14] Rappaport, T.S., Characterization of UHF multipath radio channels in factory buildings, *IEEE Transactions on Antennas and Propagation*, 37(8), pp.1058–1069, August 1989.

[15] Rappaport, T.S., Seidel, S.Y. & Takamizawa, K., Statistical channel impulse response models for factory and open plan building radio communication system design. *IEEE Transactions on Communications*, 39(5), pp. 794–807, May 1991.

[16] Saleh, A.A.M. & Valenzuela, R.A., A statistical model for indoor multipath propagation. IEEE *Journal on Selected Areas in Communications*, 5(2), pp. 128–137, February 1987.

[17] Proakis, J.G., *Digital Communications*, McGraw-Hill International Edition, Fourth Edition, 2001.

[18] Rappaport, T.S., *Wireless Communications: Principles and Practice*, 2nd Edition, 31st Dec, 2001.

[19] Anatory, J., Theethayi, N., Thottappillil, R., Kissaka, M.M. & Mvungi, N.H., Broadband power line communications: the channel capacity analysis. *IEEE Transactions on Power Delivery*, 23(1), pp. 164–170, January 2008.

[20] Paul, C.R., *Analysis of Multi-conductor Transmission Lines*, Wiley Series in Microwave and Optical Engineering, Kai Chang, Series, 1994.

[21] Vines, R.M., Trissell, H.J., Gale, L.J. & Ben O'Neal J., Noise on residential power distribution circuit. *IEEE Transactions on Electromagnetic Compatibility*, EMC-26, pp.161–168, November 1984.

[22] Chen, W.Y., *Home Networking Basis: Transmission Environments and Wired/ Wireless Protocols*, Prentice Hall, Professional Technical Reference Upper Saddle River, NJ 07458, pp. 103–143, 2004.

[23] Liu, D., Flint, E., Gaucher, B. & Kwark, Y., Wideband AC powerline characterization. *IEEE Transactions on Consumer Electronics*, 45(4), pp. 1087–1097, November 1999.

[24] Amirshahi, P. & Kavehrad, M., High-frequency characteristics of overhead multiconductor power lines for broadband communications. *IEEE Journal on Selected Areas in Communications*, 24(7), pp. 1292–1303, July 2006.

CHAPTER 6

Modulation and coding techniques for power-line communication systems

1 Introduction

There are various modulation schemes which have been studied for their suitability in broadband power-line communication (BPLC) applications [1–4]. Those modulation methods are orthogonal frequency division multiplexing (OFDM), spread spectrum modulation, multi-carrier-code division multiple access (MC-CDMA), discrete multitone modulation (DMT) and quadrature amplitude modulation (QAM) [4].

The adopted modulation scheme should provide low bit error rates (BERs) at low received signal to noise ratios (SNRs), thereby performing well in fading conditions (power-line case), occupies a minimum bandwidth, and overall making it simple and cost-effective to implement. The performance of a modulation scheme is often measured in terms of its power and bandwidth efficiencies. In digital communication systems, in order to increase noise immunity, it is necessary to increase the signal power. However, the amount by which the signal power could be increased to obtain a certain level of fidelity (an acceptable BER) depends on the particular type of modulation employed. Power-line channels are affected by interference and multipath phenomenon; hence a single carrier modulation is not advisable.

2 Orthogonal frequency division multiplexing

A power-line channel is characterized by a multipath fading environment similar to wireless networks. This is due to a number of concentrated (distributed) branches and different connected load impedances, including line length (both direct and branched lengths) as discussed in previous chapters. The delayed signals due to connected loads (i.e. refrigerators, computer power supplies, transformers, etc.) and branches (either concentrated at a node or distributed) interfere with the direct waves and cause inter-symbol interference (ISI), which degrades the network performance. Because the delayed waves interfere with the direct waves and degrade the systems, the delay must be eliminated as far as possible; the only means is

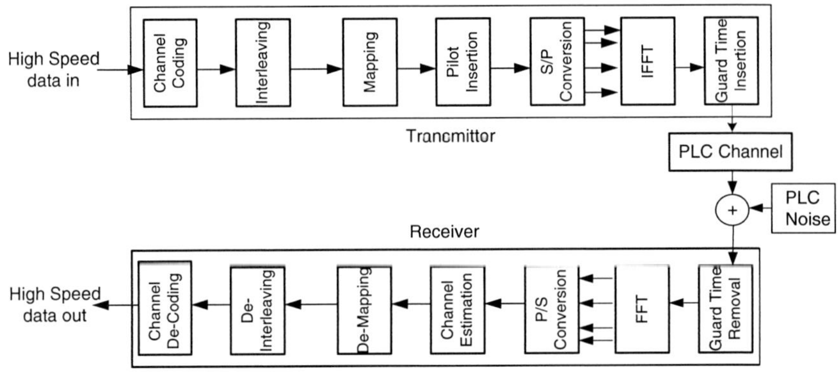

Figure 1: General configuration/scheme of an OFDM transmission system.

to use equalization techniques. However, achieving equalization at megabits per second is more cumbersome. OFDM is based on parallel transmission broadband data which reduces the effects of multipaths and leads to unnecessary equalization techniques [5–7].

The general configuration of an OFDM transmission system is shown in Fig. 1, where the transmitted high-speed data is first coded and interleaved and then mapped. Afterwards, the data are distributed as parallel data transmission in several channels, in which the transmitted high-speed data is converted into slow parallel ones in several channels. Increasing the number of parallel transmission channels reduces the data rate that each individual sub-channel must convey. The transmitted data of each parallel sub-channel is modulated by either M-ary phase shift keying (PSK) or M-ary QAM (M-QAM). The data are fed into an inverse fast Fourier transform (IFFT) circuit and then the OFDM signal is generated. The signal is fed into a guard time insertion circuit to reduce ISI then into a power-line communication (PLC) channel. At the receiver the guard time is removed, and the orthogonality of channels can be maintained by using the FFT circuit at the receiver. Because the data in a FFT circuit are parallel, a parallel to serial conversion is needed and since the power line uses a coherent detection system, channel estimates are necessary. The estimates are important so that data can be demodulated correctly. The performance of a modulation scheme in any communication channel can be determined through BER performance [6, 7].

The estimates are necessary so that data can be demodulated correctly. Generally, in a power-line channel environment, the received signal is the linear combination of several transmitted signals $s(t)$. The generalized expression for the received signal is as in (1). The parameters $h_i\ (t)$, $\tau_i\ (t)$, $s_l\ (t)$ and $z(t)$ are the complex attenuation factor of the received path i, time delay of path i, equivalent low-pass of the OFDM transmitted signal and noise in the power-line networks, respectively [7, 8]. Figure 2a shows the transmission configuration of an OFDM power-line channel. In Fig. 2a, the parameters $b_0 = e^{j2\pi f_0 t}$, $b_1 = e^{j2\pi f_1 t}$, $b_{N-2} = e^{j2\pi f_{N-2} t}$ and $b_{N-1} = e^{j2\pi f_{N-1} t}$. Figure 2b shows the receiver configuration of an OFDM system.

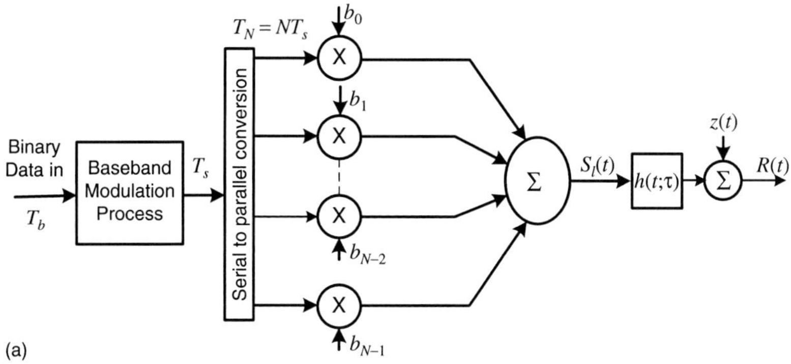

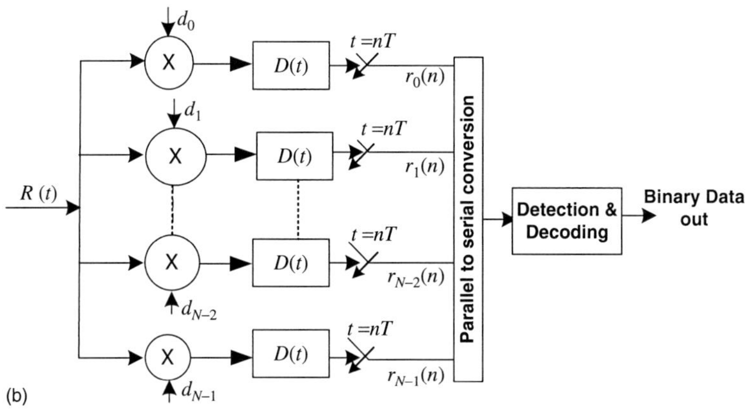

Figure 2: (a) Transmitter and (b) receiver configuration for OFDM signals with power-line channel and noise.

In Fig. 2b, the parameters $d_0 = e^{-j2\pi f_0 t}$, $d_1 = e^{-j2\pi f_{N-2} t}$, $d_{N-2} = e^{-j2\pi f_1 t}$, $d_{N-1} = e^{-j2\pi f_{N-1} t}$ and $D(t) = \frac{1}{\sqrt{T}} \int_T (\cdot) dt$.

$$R(t) = \sum_{i=0}^{L-1} A_i(t) s(t - \tau_i(t)) + n(t) \tag{1}$$

$$R(t) = \sum_{i=0}^{L-1} h_i(t) s_l(t - \tau_i(t)) + z(t) \tag{2}$$

$$h_i(t) = A_i(t) e^{-j2\pi f_c \tau_i(t)} \tag{3}$$

$$T = NT_S >> T_m = \frac{1}{B_m} \tag{4}$$

For a given transmission high data rate, the OFDM block duration must be much longer than the maximum delay spread of channel T_m with a proper selection of number of multi-carrier N. The parameter B_m is the coherence bandwidth of the power-line multipath channel. The channel impulse response is given as in (5) and it can be expanded to appear as in (6). The corresponding sub-channel response is given in (7). Assuming power-line channels with impulse responses as in (8) and (10), the corresponding sub-channel responses are as in (9) and (11), respectively [6].

$$h(t;\tau)=\sum_{i=0}^{L-1} h_i\delta(t-\tau_i) \tag{5}$$

$$h(t;\tau)=h_0\delta(t)+h_1\delta(t-\tau_1)+h_2\delta(t-\tau_2)+h_3\delta(t-\tau_3)+\cdots+h_{L-1}\delta(t-\tau_{L-1}) \tag{6}$$

$$H_k=h_1+h_2e^{-\pi\frac{k}{N}}+h_3e^{-\pi\frac{2k}{N}}+h_4e^{-\pi\frac{3k}{N}}+h_5e^{-\pi\frac{4k}{N}}+h_6e^{-\pi\frac{5k}{N}}+\cdots \tag{7}$$

$$h(t)=0.7\delta(t)-0.25\delta(t-0.5T_b)+0.7\delta(t-1.5T_b)+0.5\delta(t-3T_b) \tag{8}$$

$$H_k=0.7-0.25e^{-\pi\frac{k}{N}}+0.7e^{-\pi\frac{3k}{N}}+0.5e^{-\frac{6k}{N}} \tag{9}$$

$$h(\tau)=0.8\delta(\tau)-0.6\delta(\tau-1.5T)+0.3\delta(\tau-4T) \tag{10}$$

$$H_k=0.8-0.6e^{-\pi\frac{3k}{N}}+0.3e^{\frac{-8k}{N}} \tag{11}$$

The implementation of an OFDM system can be done either using binary phase shift keying (BPSK) or quadrature phase shift keying (QPSK) or M-ary quadrature amplitude modulation (M-QAM). The BER performance of a PLC-OFDM based system using BPSK, QPSK and M-QAM is given by (12)–(14), respectively [5–8]. The parameters E_b, N_m, H_k and N are the energy of the signal, noise power, sub-channel response and number of sub-channels, respectively. In (15), the parameters T_N and T_{guard} are information time and guard time of the OFDM symbol [5–8] as shown in Fig. 3.

$$P_{bk}=\frac{1}{N}\sum_{k=0}^{N-1}Q\left(\sqrt{\frac{2|H_k|^2 a_g E_b}{N_m}}\right) \tag{12}$$

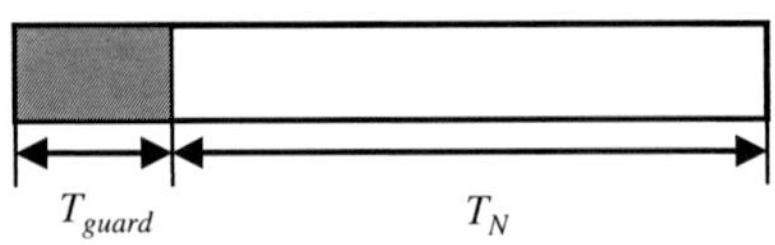

Figure 3: OFDM frame with guard time and symbol duration.

$$P_{bk} = \frac{1}{N}\sum_{k=0}^{N-1} 2\cdot Q\left(\sqrt{2\frac{|H_k|^2 a_g E_b}{N_m}}\right) - Q^2\left(\sqrt{2\frac{|H_k|^2 a_g E_b}{N_m}}\right) \tag{13}$$

$$P_{bk} \approx \frac{1}{N}\sum_{k=0}^{N-1} 4\left(1-\frac{1}{\sqrt{M}}\right)Q\left(\sqrt{\frac{3\log_2(M)|H_k|^2 a_g}{M-1}\frac{E_b}{N_m}}\right) \tag{14}$$

$$a_g = \frac{T_N}{T_N + T_{\text{guard}}} \tag{15}$$

3 Spread spectrum modulation

The direct sequence spread spectrum modulation scheme uses a technique where the spreading waveform has much higher frequency than the desired data rate. Assume R is the data rate of the signal with pulse $T = 1/R$ and that the spreading is transmitted at the rate R_c, the increase in data rate is R_c/R. The spreading transmitted rate R_c is known as the chip rate. The width of each pulse in the modulating sequence T_c is known as chip duration. Figure 4 shows the block diagram of a spread spectrum system [5–7]. Consider a power-line channel as a multipath channel with a coherence bandwidth B_m and channel impulse response $h(\tau,t)$ as in (16), which is an inverse of the power-line channel frequency response $H(f)$. The parameter f_c is the carrier frequency, $A_i(t)$ and $\tau_i(t)$ are the attenuation coefficient and path delay of path i, respectively. The received direct sequence spread spectrum modulation signal can be represented as in (17). The parameter h_i is the attenuation factor of path i and L is the resolvable multipath [8].

$$h(\tau,t) = \sum_{i=-\infty}^{+\infty} A_i(t)e^{-j2\pi f_c \tau_i(t)}\delta(\tau - \tau_i(t)) \tag{16}$$

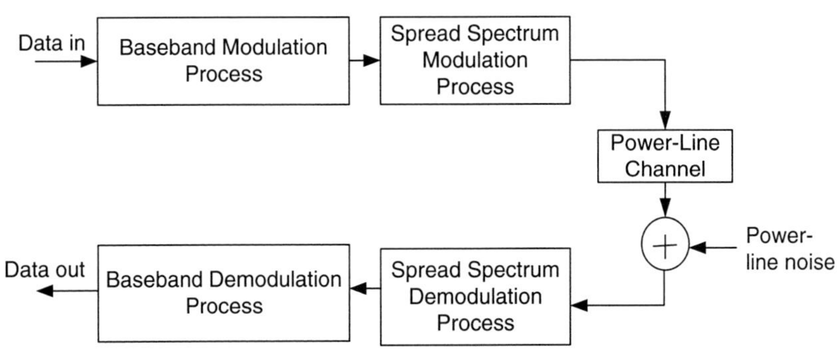

Figure 4: Block diagram of a BPLC system based spread spectrum as a modulation scheme.

$$R_r(t) = \sum_{i=0}^{L} h_i s_d(t - \tau_i) c(t - \tau_i) + z(t) \tag{17}$$

In the PLC system, for random spreading codes, the output of the BPSK correlation receiver can be approximated as in (18). The parameter z_n is a complex noise component in the power-line channel, which is variable with zero mean, and the variance is as shown in (19). $W_s = 1/T_s$ and $R_b = 1/T_b$, T_s is the symbol duration and T_b is the information symbol rate. Assume that the power-line channel is a fading channel such that each path is Rayleigh distributed with probability density function as in (20). The parameter $2\sigma_i^2$ is the average power of path i. The average BER of a direct sequence spread spectrum signal with coherent BPSK modulation in a power-line channel when a RAKE receiver is used can be upper bounded as in (21) [6, 7].

$$V_n = \left(\sum_{i=0}^{L-1} |h_i|^2 \right) \sqrt{E_b} a_n + z_n \tag{18}$$

$$\sigma_z^2 = (N_0 + E_b R_b / W_s) \sum_{i=0}^{L-1} |h_i|^2 \tag{19}$$

$$pA_i(a) = \frac{a}{\sigma_i^2} e^{-a^2/2\sigma_i^2}, \quad a \geq 0,\ i = 0, 1, \ldots, L-1 \tag{20}$$

$$P_b \leq \frac{1}{2} \prod_{i=0}^{L-1} \left(\frac{1}{1 + \dfrac{2\sigma_i^2 E_b}{\sigma_z^2}} \right) \tag{21}$$

4 Multi-carrier spread spectrum modulation

BPLC systems based on the multi-carrier spread spectrum (MC-SS) modulation scheme are combinations of direct sequence spread spectrum and OFDM modulation. The idea behind MC-SS is to spread the information symbol over different frequencies by transmitting each chip over different frequencies. The baseband modulator can be either (PSK or QAM or phase amplitude modulation (PAM) [9]. Since MC-SS is based on OFDM, it is quite effective in multipath channels like power-line channels as it has the capability to resolve all the diversity provided by the channel. Considering a power-line channel with impulse response $h_i(t)$ and τ_i the time delay of path i, the multipath spread spectrum received signal is given as in (22). Figure 5a shows the general configuration of an MC-SS transmitter system [6, 7]. In Fig. 5a, the parameters $d_0 = e^{j2\pi f_0 t}$, $d_1 = e^{j2\pi f_1 t}$, $d_{N-2} = e^{j2\pi f_{N-2} t}$ and $d_{N-1} = e^{j2\pi f_{N-1} t}$.

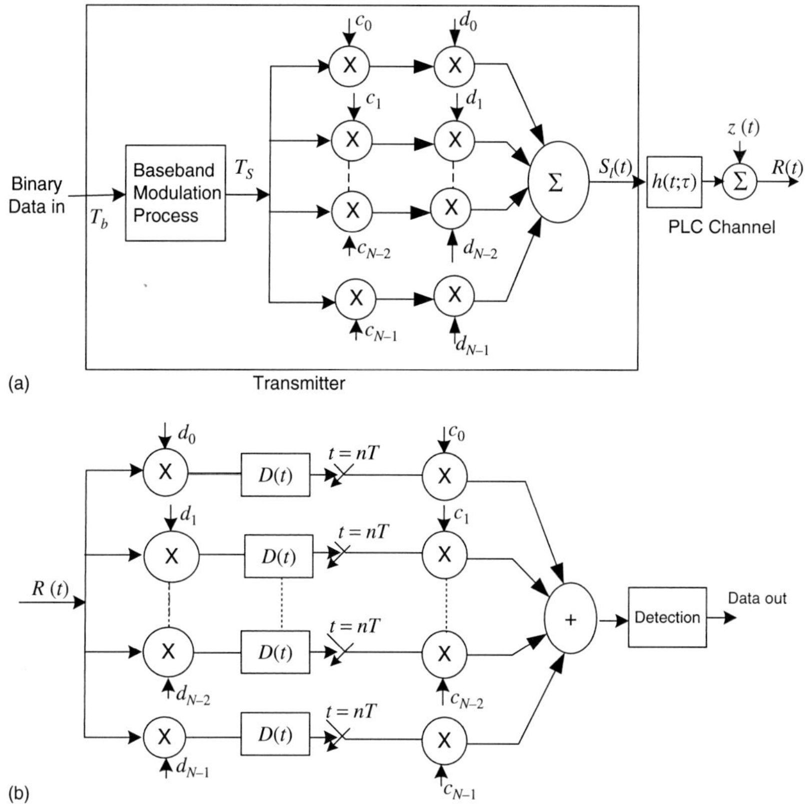

Figure 5: General configuration of a multi-carrier spread spectrum (a) transmitter and (b) receiver with power-line channel.

$$R(t) = \sum_{i=0}^{L-1} h_i s(t - \tau_i) + z(t) \tag{22}$$

where $s(t)$ and $z(t)$ are the transmitted signal and noise in the power-line network (which is complex in nature), respectively. Figure 5b shows the general configuration of an MC-SS receiver system [6, 7]. The parameters $d_0 = e^{-j2\pi f_0 t}$, $d_1 = e^{-j2\pi f_1 t}$, $d_{N-2} = e^{-j2\pi f_{N-2} t}$, $d_{N-1} = e^{-j2\pi f_{N-1} t}$ and $D(t) = \int_T (\cdot) dt$.

$$y_k(n) = H_k(n) s(n) c_k + z_k(n) \tag{23}$$

Assuming that the guard interval is larger compared to the maximum delay spread of a power-line channel, the output sample of the correlation receiver, during the nth symbol interval can be written as in (23), where $k = 0, 1, \ldots, N-1$. The parameter $z(n)$ is a complex variable representing noise in the power-line channel, $s(n)$ is the baseband modulated symbol during the nth symbol interval and $H_k(n)$ is a sampled power-line channel transfer function at sub-carrier k, and is given by (24), and since the sampled noise is a complex random variable, optimum signal dispreading is obtained as in (25), which can be expanded as in (26). The SNR per symbol is directly obtained from (26) and can be represented as in (27). The parameter c_k is a spreading factor/element [6, 7].

$$H_k(n) = \sum_{i=0}^{L-1} h_i e^{-j2\pi k\tau_i/T_s} \tag{24}$$

$$y(n) = \sum H_k^*(n) c_k^* y_k(n) \tag{25}$$

$$y(n) = \left(\frac{1}{N} \sum_{k=0}^{N-1} |H_k(n)|^2 \right) s(n) + \sum_{k=0}^{N-1} H_k^*(n) c_k^* z_k(n) \tag{26}$$

$$SNR = \left(\frac{1}{N} \sum_{k=0}^{N-1} |H_k(n)|^2 \right) \frac{E_s}{N_o} \tag{27}$$

The performance indication of the modulation scheme in a power-line channel is through BER performance. The BER performance of BPLC-based MC-SS as a modulation scheme is given by (28) [6]. The parameters E_s, N_m, H_k, and N are the energy of the signal, noise power, sub-channel response and number of sub-carriers, respectively.

$$P_{bk} = \sqrt{2 * \left(\frac{1}{N} \sum_{k=0}^{N-1} |H_k(n)|^2 \right) \frac{E_s}{N_m}} \tag{28}$$

5 Discrete multitone modulation

DMT is a discrete time approximation to the multi-carrier modulation system. It uses the discrete Fourier transform (DFT) and inverse discrete Fourier transform (IDFT) to demodulate and modulate the data to be transmitted, respectively. The transformations can be efficiently implemented with digital signal processing (DSP) using FFT. In DMT, the sub-carrier is sometimes referred to as tones which are modulated in a real discrete time signal by IFFT operation. Figure 6 shows a block diagram of a PLC system based on a DMT transmitter and receiver. It can be observed that the data are first coded by forward error correction (FEC) then constellation, mapped and shuffled. The shuffled data are subjected to IFFT operation and a cyclic prefix is inserted to overcome ISI. The data are converted from digital to analogue (D/A) format, then filtered and transmitted in the power-line channel. The transmitted data are received at the receiving filter; then analogue to digital (A/D) operation is performed and the cyclic prefix is removed. The data are subjected to an FFT operation; then constellation, de-mapping and tone shuffle are performed. Finally, the received data are checked for errors and corrected using the FEC decoder.

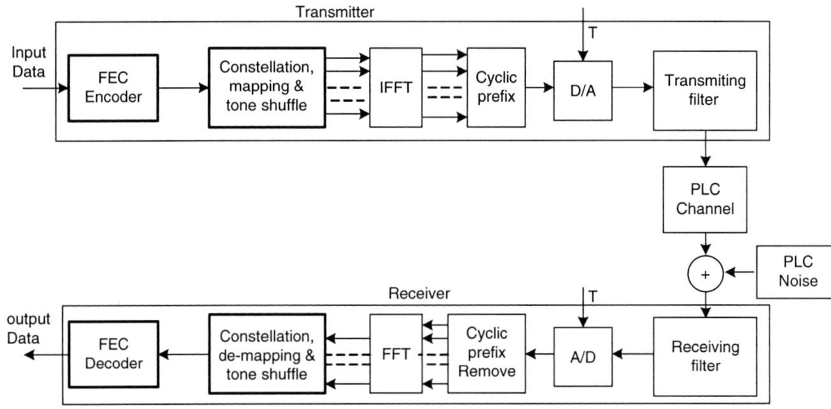

Figure 6: Block diagram of a discrete multitone transmitter and receiver.

Normally the performance of a single carrier can be extended to multi-carriers systems such as DMT. For a single-carrier QAM on ideal additive white Gaussian noise (AWGN) channels, the union bound estimates for the probability of a symbol error is as shown in (29). For a PLC system with channel transfer function $H_k(f)$, an un-coded system requires an SNR gap of approximately 9.8 dB to operate at a symbol error probability of 10^{-7}. The coding scheme in such a system is used to reduce the minimum distance necessary to achieve a desired BER. The system may also include an error margin, a safety factor included to protect the system channel performance in case of any channel degradation. In DMT PLC systems, in terms of performance the SNR gap is the measure that accounts for both the coding gain and the error margin in performance analysis. The union bound for the probability of error is as in (30) in terms of the SNR gap [9–12]. Note that the gap is defined as in (31).

$$P_e \leq 4Q\left(\frac{d_{\min}}{2\sigma}\right) \tag{29}$$

$$P_e \leq 4Q(\sqrt{3\Gamma}) \tag{30}$$

$$3\Gamma = \frac{|H_k|^2 d_k^2}{4\sigma_k^2} \tag{31}$$

For a system with error margin γ_{mg} and code with coding gain γ_{cg} if the uncoded system requires an SNR gap of γ_g to achieve the target error probability, the gap necessary with coding and error margin is as in (32).

For a DMT PLC system like any other multi-carrier PLC system with N sub-channels, each consisting of QAM signals with distances d_k, average energies E_k, power-line channel transfer function gain H_k and noise variation σ_k, the channel signal to noise ratio SNR_i is given as in (33), the symbols transmitted b_k is given as in (34) and the total number of bits transported in one multi-carrier symbol is as in (35). [12].

$$\Gamma = \Gamma_g + \gamma_{mg} - \gamma_{cg} \tag{32}$$

$$SNR_k = \frac{|H_k|^2 E_k}{2\sigma_k^2} \tag{33}$$

$$b_k = \log_2\left(1 + \frac{SNR_k}{\Gamma}\right) \tag{34}$$

$$b = \sum_{i=1}^{N} \log_2\left(1 + \frac{SNR_k}{\Gamma}\right) \tag{35}$$

6 Coding techniques for BPLC systems

A power-line network has a noisy environment. Hence, to be able to receive data appropriately, channel coding is very important. Channel coding protects data from errors by introducing redundancies in the transmitted data. Channel codes that are used to detect errors are called error detection codes, while codes that can detect and correct errors are called error correction codes. The basic purpose of introducing redundancies in the data is to improve power-line links performance. The introduction of redundant bits increases the raw data used in the link and hence increases the bandwidth requirement for a fixed source rate in the system. This reduces the bandwidth efficiency of the link in high SNR conditions but provides excellent BER performance at low SNR values. In the communication, there are two different types of error correcting and detection codes. These include block codes and convolution codes. Convolution codes are good for error correction in the communication systems, while block codes are good for error detection and, in some cases, corrections. Like in any other communication system, coding can be used to improve the performance of a PLC system when other means of improvement such as increasing transmitter power or using a more sophisticated demodulator are impractical [12–18].

6.1 Convolutional codes

In convolutional codes, a binary encoder takes a stream of information bits and converts it into a stream of transmitted bits, using a shift register bank. Redundancy for recovery from communication channel errors is provided by transmitting more bits per unit time than the number of information bits transmitted per unit time. The maximum likelihood decoding is normally done using the Viterbi algorithm. Also, there are other algorithms which are commonly used such as the Bahl–Cocke–Jelinek–Raviv algorithm and soft output Viterbi algorithm. In some cases, in power-line environments which experience burst errors, convolutional codes are used as inner codes with block codes as outer codes to form concatenated codes. For a code with higher correcting capability such as PLC, the outer code is used to recover from such burst error patterns in the decoding of the inner code [15, 17–19].

6.2 Error probabilities for convolutional codes

It has been pointed out previously that the performance of PLC systems like any communication system can be obtained through probability error (BER). For convolutional codes, there are two error probabilities associated with it, namely first event and bit error probabilities. The first event error probability, P_e, is the probability that an error begins at a particular time and is obtained as in (36). The bit error probability, P_b, is the average number of bit errors in the decoded sequence and is represented in (37). Usually, these error probabilities are defined using the Chernoff bounds and are derived in [6, 7, 16–18]. For BPSK modulation p can be obtained as in (38), and the $Q(x)$ function is represented in (39).

$$P_e < T(D,N,J)_{D=\sqrt{4p(1-p)},N=1,J=1} \tag{36}$$

$$P_b < \left.\frac{T(D,N,J)}{dT}\right|_{D=\sqrt{4p(1-p)},N=1,J=1} \tag{37}$$

$$p = Q\left(\sqrt{\frac{2rE_b}{N_o}}\right) \tag{38}$$

$$Q(x) = \int_x^\infty \frac{1}{\sqrt{2\pi}} e^{-u^2/2} du \tag{39}$$

For soft-decision decoding, the first event error and bit error probabilities are defined as in (40) and (41), respectively.

$$P_e < T\left(D,N,J\right)_{D=e^{-rE_b/N_o},N=1,J=1} \tag{40}$$

$$P_b < \left.\frac{T\left(D,N,J\right)}{dT}\right|_{D=e^{-rE_b/N_o},N=1,J=1} \tag{41}$$

For the entire PLC system which consists of convolutional encoder with biterbi decoder, the probability error $\overline{P}_b$ is presented as in (42) [9, 16–18]. The parameters b, d_{free} and R_c are the puncture period, free distance of convolutional code and code rate, respectively. β_d is the total number of bit errors that occur in all the incorrect paths in the trellis, which differs from the correct path in exactly d positions. In (42), P_2 (d) is the probability of choosing an incorrect path that differs from the correct path in exactly d positions. For coded OFDM systems the probability error is given (43). In (43) the entire expression is divided by N which implies that the sub-carriers are averaged. The upper bound in (43) can be expressed in a slightly different form if the Q-function is upper-bounded by an exponential such as $Q(\sqrt{d\gamma_{bk}}) \le e^{-\gamma_{bk}d} = D^d\big|_{D=e^{-\gamma_{bk}}}$ [17]. The upper bound for a soft-decision first event probability can be expressed by (46). The generating transfer function $T(D,B)$ with d_{free} can be obtained from Viterbi [16, 17] and is given as in (47).

$$\bar{P}_b < \frac{1}{b}\sum_{d=d_{\text{free}}}^{\infty} \beta_d P_2(d) \tag{42}$$

$$P_2(d) = \frac{1}{N}\sum_{k=0}^{N-1} Q\left(\sqrt{d\gamma_{bk}}\right) \tag{43}$$

$$\gamma_{bk} = \frac{2R_c\left|H_k\right|^2 \alpha_g E_b}{N_m} \tag{44}$$

$$\bar{P}_{bk} < \frac{1}{2}\frac{1}{b}\frac{1}{N}\sum_{k=0}^{N-1}\left(\sum_{d=d_{\text{free}}}^{\infty} \beta_d Q(d\gamma_{bk})\right) \tag{45}$$

$$\bar{P}_{bk} < \frac{1}{2}\frac{1}{b}\frac{1}{N}\sum_{k=0}^{N-1}\left(\frac{dT(D,B)}{dB}\right)\Bigg|_{B=1,D=e^{-\gamma_{bk}}} \tag{46}$$

$$T(D,B) = \frac{D^{d_{\text{free}}}B}{1-2DB} \tag{47}$$

6.3 Block codes

The block codes like convolutional codes are also applicable in power-line channel environments. The block codes consist of a set of independent fixed length sequences called code words. The length of a code word is the number of coded symbols (bits) in the sequence and is denoted by n. Examples of block codes which could be applied in power-line channels are Hamming codes, Hadamard codes, Golay codes, cyclic codes, Bose–Chaudhuri–Hacknguem (BCH) codes and Reed–Solomon codes. Other codes include turbo code and concatenated codes and current research has proposed low-density parity-check (LDPC) code and tornado codes, which is a combination of LDPC and fountain codes [13–16]. Other codes include online codes, luby transform (LT) codes and Raptor codes. Within these block codes there are two types of decision decoding mechanisms which include hard decision and soft decision decoding [16–24].

6.4 Error probabilities for block codes

For block codes the code word error probability is upper-bounded by (48) where the parameter $t = [(d_m - 1)/2]$ and the equality holds when the linear block code is perfect. In general, for low values of the channel error probability p, the upper bound is quite close to the exact value. A very simple and useful approximation of the average bit error probability can be derived from the fact that, given a decoding failure, the most probable result is an adjacent code word of distance d_m, i.e. containing a total of d_m errors. With this in mind the approximation of bit error probability is as shown in (49). Then the bit error probability for block codes are upper-bounded as in (50) [8, 9, 16–18]. Note that the channel error probability p will depend on the modulation scheme used.

$$P_{cw} < \sum_{k=t+1}^{n} \binom{n}{k} p^k (1-p)^{n-k} = 1 - \sum_{k=0}^{t} \binom{n}{k} p^k (1-p)^{n-k} \tag{48}$$

$$P_b \approx \frac{d_m}{n} P_{cw} \tag{49}$$

$$P_b < \sum_{i=t+1}^{n} \frac{i+t}{n} \binom{n}{i} p^i (1-p)^{n-i} \tag{50}$$

6.5 Concatenated codes

Concatenation is a method of building long codes out of shorter ones with the aim of attempting to meet the problem of decoding complexity. This can be achieved by breaking the required computation into manageable segments. The concatenated Reed–Solomon (RS) codes/interleaved Viterbi channel coding is used due its effectiveness for burst-error correction [25–36], which is the case in power-line networks as found in other applications and proposed by Home Plug [34]. The concatenated coding scheme consists of an outer block code over GF(2^B) and an inner binary convolutional code. Assuming that interleaving between the Viterbi decoder and the RS decoder is sufficiently long to break up long bursts of errors out of the Viterbi decoder, the RS symbol error probability (P_B) for symbols in GF(2^B) can be upper-bounded by the simple union bound as in (51). The parameter $\overline{P}_b$ is the probability error at the output of the Viterbi decoder, which can be expressed by (52) [30–34]. The parameters d_{free} and R_c are the free distance of convolutional code and code rate, respectively. β_d is the total number of bit errors that occur in all the incorrect paths in the trellis, which differs from the correct path in exactly d positions. $P_2(d)$ is the probability of choosing an incorrect path that differs from the correct path in exactly d positions. Note that in (53), the expression is for PLC based on OFDM system, where H_k and N are the power-line channel response and number of sub-carriers, respectively. The total error probability of an RS code accounts for both decoder failure probability and decoder error probability. The symbol-error probability is given by (54). For a given transmitted signal and if there are $(2^B - 1)$ erroneous signals with equally likely bits, then the probability of bit given that the received signal is in error is equal to $(2^{B-1})/(2^B - 1)$ [31, 32]. Then the total probability error of concatenated codes is given by (55) [33–35].

$$P_B \leq B\overline{P}_b \tag{51}$$

$$\overline{P}_b < R_c \sum_{d=d_{\text{free}}}^{\infty} \beta_d P_2(d) \tag{52}$$

$$P_2(d) = \frac{1}{N}\sum_{k=0}^{N-1} Q\left(\sqrt{\frac{2R_c\left|H_k\right|^2 a_g E_b}{N_m}}\right) \tag{53}$$

$$P_w \leq \frac{1}{n_2}\sum_{i=t+1}^{n_2} i\binom{n_2}{i} P_B{}^{i}(1-P_B)^{n_2-i} \tag{54}$$

$$P_{CD} \leq \frac{1}{n_2}\left(\frac{2^{B-1}}{2^B-1}\right)\sum_{i=t+1}^{n_2} i\binom{n_2}{i} P_B{}^{i}(1-P_B)^{n_2-i} \tag{55}$$

References

[1] Hrasnica, H., Reservation MAC protocols for power-line communications. *International Symposium on Power-line Communications and Its Applications*, Malmo, Sweden, 4–6 April 2001.

[2] Lutz, H.L. & Huber, J.B., Bandwidth efficient power-line communications based on OFDM. *International Journal of Electronics and Communications Archive für Elektronik und übertragungstechnik Special Issue: Power-line Communications*, **54**(1), February 2000.

[3] Langfeld, P.J., Zimmermann, M. & Dostert, K., Power-line communication system design strategies for local loop access. *Proceedings of the Workshop Kommunikationstechnik*, Technical Report ITUU-TR-1999/02, ISSN1433-9080, pp. 21–26, July 1999.

[4] Langfeld, P.J., The capacity of typical power-line reference channels and strategies for system design. *Proceedings of the 5th International Symposium on Power-Line Communications and its Applications, ISPLC-2001*, Malmö, Sweden, pp. 271–278, 4–6 April 2001.

[5] Prasad, R., *OFDM for Wireless Communications Systems*, Artech House: Boston, MA, 2004.

[6] Ahlin, L., Zander, J. & Slimane, B., *Principles of Wireless Communications*, Narayana Press: Denmark, 2006.

[7] Goldsmith, A., *Wireless Communications*, Cambridge University Press, 2005.

[8] Ma, Y., So, H.P.L. & Gunawan, E. Performance analysis of OFDM systems for broadband power line communications. *IEEE Transactions on Power Delivery*, **23**(1), pp. 132–139, 2008.

[9] Lapidoth, A., On the probability of symbol error in Viterbi decoders. *IEEE Transactions on Communications*, **45**(2), pp. 152–155, 1997.

[10] Bingham, J.A.C., Multicarrier modulation for data transmission: an idea whose time has come. *IEEE Communications Magazine*, pp. 5–14, May 1990.

[11] Cioffi J., A multicarrier primer. ANSI T1E1.4 Committee Contribution 91-157, Boca Raton, November 1991.

[12] Bingham, J., RFI suppression in multicarrier transmission systems. *Proceedings of IEEE Globe Communication Conf.*, 2, pp. 1026–1030, 1996.

[13] Carlson, B.A., *Communication Systems: An Introduction to Signals and Noise in Electrical Communication*, McGraw Hill International Editions, Electrical & Electronic Engineering Series, 3rd edn, pp. 469–508, 1986.

[14] Richard, V.N. & Prasad, R., *OFDM for Wireless Multimedia Communications*, Artech House: Boston, 2000.

[15] Shanmugam, S.K., *Digital and Analog Communication Systems*, Wiley, 2006.

[16] Proakis, J.G., *Digital Communications*, McGraw-Hill International Edition, 4th edn, 2001.

[17] Viterbi, A.J., Convolutional codes and their performance in communication systems. *IEEE Transaction on Communication*, COM-19, pp. 751–772, October 1971.

[18] Moon, KT., *Error Correction Coding*, John Wiley & Sons: New Jersey, ISBN 978-0-471-64800-0, 2005.

[19] MacKay, D.J.C., *Information Theory, Inference and Learning Algorithms*, Cambridge University Press, ISBN 0-521-64298-1, September 2003.

[20] Bhattacharryya, D.K. & Nandi, S., An efficient class of SEC-DED-AUED codes. *International Symposium on Parallel Architectures, Algorithms and Networks, ISPAN '97*, pp. 410–415, 1997.

[21] Thompson, T.M., *From Error Correcting Codes through Sphere Packings to Simple Groups*, Carus Mathematical Monographs, Mathematical Association of America, 1983.

[22] Curtis, R.T. A new combinatorial approach to M_{24}. *Math. Proc. Camb. Phil. Soc.*, **79**, pp. 25–42. 1976

[23] Lin, S. & Costello, D., *Error Control Coding: Fundamentals and Applications*, Prentice-Hall: Englewood Cliffs, NJ, 2004.

[24] Gilbert, W.J. & Nicholson, W.K., *Modern Algebra with Applications*, 2nd edn, Wiley, 2004.

[25] Lidl, R. & Pilz, G., *Applied Abstract Algebra*, 2nd edn, Wiley, 1999.

[26] MacWilliams, F.J. & Sloane, N.J.A., *The Theory of Error-Correcting Code*, North-Holland Publishing Company: New York, 1977.

[27] Reed, I.S., & Chen, X., *Error-Control Coding for Data Networks*, Kluwer Academic Publishers: Boston, 1999.

[28] Reed, I.S. and Solomon, G., Polynomial codes over certain finite fields. *SIAM Journal of Applied Math.*, **8**, pp. 300–304, 1960.

[29] Liu, K.Y. & Lee, J., Recent results on the use of concatenated Reed-Solomon/Viterbi channel coding and data compression for space communications. *IEEE Transactions on Communications*, **32**(5), pp. 518–523, 1984.

[30] Wong, Y.F.M. & Letaief, K.B., Concatenated coding for DS/CDMA transmission with wireless communications. *IEEE Transactions on Communications*, **48**(12), pp. 1965–1969, 2000.

[31] Cideciyan, R.D., Eleftheriou, E. & Rupf, M., Concatenated Reed-Solomon/convolutional coding for data transmission in CDMA-based cellular systems. *IEEE Transactions on Communications*, **45**(10), pp. 1291–1303, 1997.

[32] Lapidoth, A., *A Foundation in Digital Communication*, University Press: Cambridge, 2009.

[33] Foerster, J. & J. Liebetreu, FEC Performance of Concatenated Reed-Solomon and Convolutional Coding with Interleaving. IEEE 802.16 Broadband Wireless Access Working Group, IEEE 802.16.1pc-00/33, June 2000.

[34] Lee, M.K., Newman, R., Latchman, H.A., Katar, S. & Yonge, L., HomePlug 1.0 powerline communication LANs – protocol description and comparative performance results. *Special Issue of the International Journal on Communication Systems on Power-line Communications*, pp. 447–473, May 2003.

[35] MacKay, D.J.C. & Neal, R.M., Near Shannon limit performance of low density parity check codes. *Electronics Letters*, July 1996

[36] Gallager, R.G., *Information Theory and Reliable Communication*, John Wiley & Sons: New York, 1968.

CHAPTER 7

Performance of PLC systems that use modulation and coding techniques

1 Noise model

Power-line channel suffers from impulsive noise interference (cause bit or burst errors in data transmission) due to connected electrical systems such as transformers, industrial switches, etc. Middleton's Class A noise model is appropriate for use in conjunction with broad band power line communications (BPLC) channel models under impulsive noise environments [1–4]. Based on the Middleton's noise model, the combination of impulsive plus background noise is a sequence of i.i.d complex random variables with the probability density function (PDF) of Class A noise as given by (1), where m is the number of impulsive noise sources and is characterized by Poisson distribution with mean parameter A called the impulsive index (2). In (2) Γ is the Gauss impulsive power ratio (GIR) which represents the ratio between the variance of Gaussian noise components σ_g^2 and the variance of impulse component σ_m^2. The variance of noise σ_z^2 is given by (3) [4]. The power-line channel and noise models have been adopted here and used to investigate the performance of medium, indoor and underground cables for power-line communication systems.

$$p_z(z) = \sum_{m=0}^{\infty} \frac{a_m}{2\pi\sigma_m^2} \exp\left(-\frac{z^2}{2\sigma_m^2} \right) \tag{1}$$

$$a_m = e^{-A}\frac{A^m}{m!}, \quad \sigma_m^2 = \sigma_g^2 \frac{\left(\frac{m}{A}\right) + \Gamma}{\Gamma} \tag{2}$$

$$\sigma_z^2 = E\left\{z^2\right\} = \frac{e^{-A}\sigma_g^2}{\Gamma} \sum_{m=0}^{\infty} \frac{A^m}{m!}\left(\frac{m}{A} + \Gamma\right) \tag{3}$$

2 Medium voltage systems

For an MV channel system, we adopt quadrature amplitude modulation (QAM) as the sub-carrier. The bit error rate performance of orthogonal frequency division

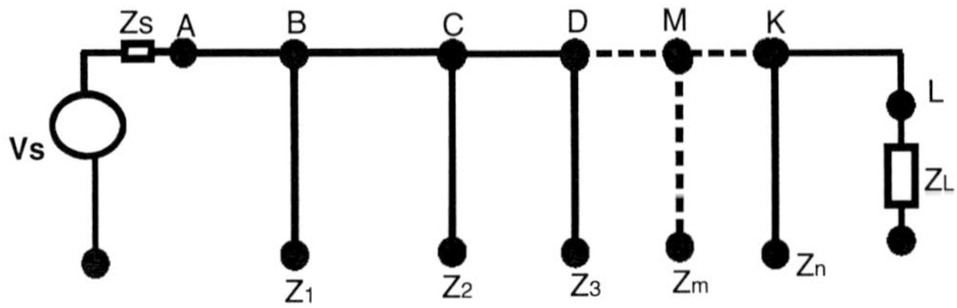

Figure 1: Power-line network with distributed branches between sending and receiving ends.

multiplexing (OFDM) system is given by (4) [5–7]. The parameters E_b, N_m, H_k, M and N are the energy of the signal, noise power, sub-channel response, modulation level and number of sub-channels, respectively. In (5), the parameters T_N and T_{guard} are information time and guard time of OFDM symbol [5].

$$P_{bk} \approx \frac{1}{N}\sum_{k=0}^{N-1} 4\left(1-\frac{1}{\sqrt{M}}\right) Q\left(\sqrt{\frac{3\log_2(M)|H_k|^2}{M-1}\frac{a_g\ E_b}{N_m}}\right) \tag{4}$$

$$a_g = \frac{T_N}{T_N + T_{guard}} \tag{5}$$

It has been observed in the previous chapters that for medium voltage channel, the maximum delay spread T_m is 4 μs. We consider an OFDM system with total frequency band B = 99.9 MHz. With such bands, a single-carrier system would have symbol time T_s of 1 ns. Considering T_m of 4 μs, there would be severe inter-symbol interference (ISI). The channel coherence bandwidth B_c is 0.25 MHz. To ensure flat fading on each sub-channel, we take BN = B/N = 0.1BC [6]. Thus number of sub-channels N needed is 3996. In the actual implementations of multi-carrier modulation, N must be a power of 2 for the discrete Fourier transformation (DFT) and inverse of DFT (IDFT) operations, in which case $N = 4096$ is appropriate. So the OFDM symbol duration is $T_N = N \times T_S = 40.96$ μs. To ensure no ISI between OFDM symbols, the length of cyclic prefix is set to $\mu = 512 > T_m/T_S$ hence, the guard interval $T_{guard} = \mu T_S = 5.12$ μs. These design parameters are used in all the cases to follow in the chapter.

2.1 Influence of number of branches

To determine the influence of branches, the power-line configuration with distributed branches as in Fig. 1 was considered. The branches between point A and L were equally distributed in the link between transmitting and receiving ends. The transmitter and receiver loads were terminated in the line characteristic impedances and the system was assumed to be synchronized. The line length between point A

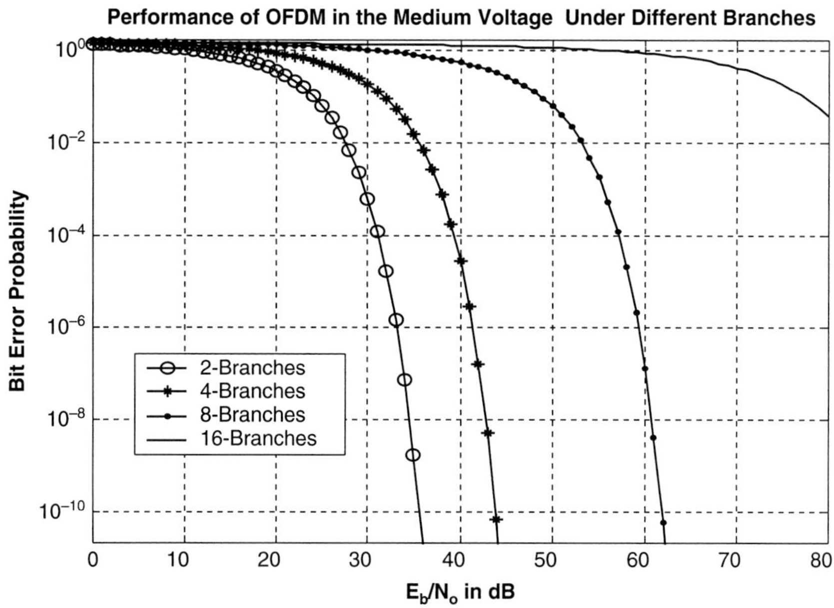

Figure 2: Simulation results for the OFDM system with 16-QAM modulation for medium voltage PLC channel for various number of branches.

and L was 1500 m, while the branch line lengths were kept at 15 m. The branches were varied as 2, 4, 8 and 16 and all branch loads were terminated in characteristic impedances. The per unit length parameters for medium voltage line is (L_e = 1.9648 μH/m, C_e = 5.6627 pF/m).

For each case the channel transfer function *H(f)* was determined and the channel was sampled. For the case of noise N_m the square root noise variance in (3) was used. In (3) values of A and GIR were 0.1 and 0.1, respectively, and *m* is taken as 3 [1]. Figure 2 shows the performance of the OFDM system for various number of branches. It can be observed that to attain a bit error probability of 10^{-10} the signal to noise ratio (SNR) per bit of 36, 45, 62, and more than 80 dB are needed for 2, 4, 8 and more than 16 branches, respectively. This indicates the average power needed to maintain sustained communication is about 4 dB/branch.

2.2 Influence of load impedances

Again, the configuration similar to Fig. 1 with four distributed branches was considered with line length between A and L as 1500 m with branch lengths as 15 m. The branches between point A and L were equally distributed in the link between transmitting and receiving ends. The transmitter and receiver loads were terminated in the line characteristic impedances and the system between transmitting and receiving ends were assumed to be synchronized.

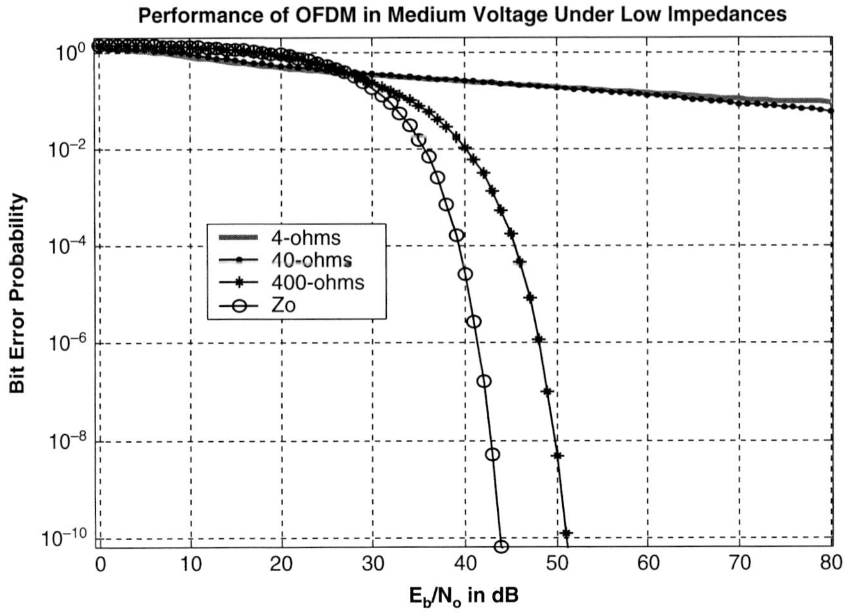

Figure 3: Simulation results for the OFDM system with 16-QAM modulation for medium voltage PLC channel for various low load branch impedances, with four distributed branches between the link.

2.2.1 Low impedance loads

We consider first the low impedance loads (loads less than branch line characteristic impedance). The load impedances at all terminals were varied as 4, 40, 400 Ω and characteristic impedance.

Figure 3 shows the performance of the OFDM system for various low load impedance cases. It is observed that the good performance can be obtained when the channel is terminated in characteristic impedances wherein the bit error probability is 10^{-10} at an SNR per bit of 45 dB. When the load impedance decreases by 200 Ω from line characteristic impedance, the power loss is about 0.0125 dB/Ω. However, as the load impedance approaches a short circuit a degraded system performance is found. This is due to the fact that at short circuit, higher deep notches exist in the system.

2.2.2 High impedance loads

We now consider the high impedance loads (impedances higher than the line characteristic impedance). The load impedances at all terminals were varied as 800 Ω, 1.6 kΩ, 3.2 kΩ and 6.4 kΩ. Figure 4 shows the performance of the OFDM system for various high impedance cases. A good channel performance is seen for 800 Ω terminations with the bit error probability of 10^{-10} at an SNR per bit of 50 dB. The power is 50, 68 and more than 80dB, for 800 Ω, 1.6 kΩ and for >3.2 kΩ,

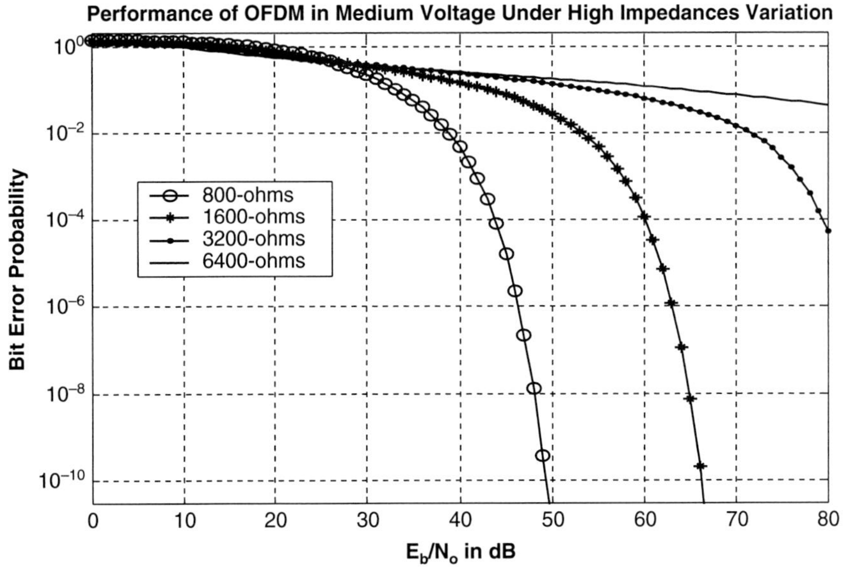

Figure 4: Simulation results for the OFDM system with 16-QAM modulation for medium voltage PLC channel for various high branch terminal impedances, with four distributed branches between the link.

respectively. If the load impedance increases above 3.2 kΩ the power loss is >80 dB indicating degraded performance (at open circuit the performance is severely degraded due to deep notches in the system).

3 Indoor systems

Consider a typical IV channel used in residences or offices [8], which connects appliances using cables having cross section 2.5 mm^2 and separation 3 mm as discussed in Ref. [8]. The transmission line parameters are L_e = 0.44388 μH/m and C_e = 61.734 pF/m. The power-line channel configuration under investigation is given in Fig. 1 with $Z_s = Z_L$ = 85 Ω. We investigate the variations in the number of branches in the link between A and L and also vary the terminal impedances connected to the branches corresponding to Fig. 1. We consider an OFDM system with total frequency band B = 99.9 MHz. With such bands, a single-carrier system would have symbol time T_s of 1 ns. The indoor power-line channel has a maximum delay spread (time span of channel impulse response) of T_m of 1 μs [8], due to this there would be severe ISI. We assumed an OFDM system with binary phase shift keying (BPSK) modulation applied to each sub-channel. The channel coherence bandwidth B_c is 1 MHz. To ensure flat fading on each sub-channel, we take BN = B/N = 0.1BC. Thus number of sub-channels N needed is 999. In the actual implementations of multi-carrier modulation, N must be a power of 2 for the DFT and IDFT operations, in which case

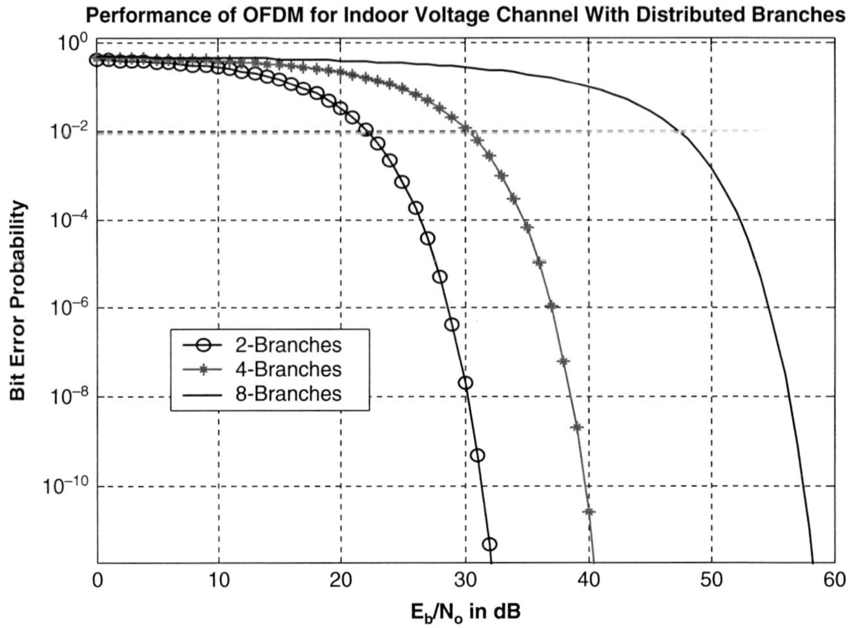

Figure 5: Simulation results for the OFDM system with BPSK modulation for indoor PLC channel for various number of branches terminated in characteristic impedances.

$N = 1024$ is appropriate. So the OFDM symbol duration is $T_N = \mathrm{N} \times T_S = 10.24$ μs. To ensure no ISI between OFDM symbols, the length of cyclic prefix is set to $\mu = 128 > T_m/T_s$; hence, the guard interval $T_{guard} = \mu T_S = 1.28$ μs. These design parameters are used in all the cases to follow in the chapter.

3.1 Influence of number of branches

To determine the influence of branches, the power-line configuration with distributed branches as shown in Fig. 1 was considered. The branches between point A and L were equally distributed between transmitting and receiving ends. The transmitter and receiver loads were terminated in the line characteristic impedances and the system was assumed to be synchronized. The line length between point A and L was 20 m, while the branch line lengths were kept at 10 m. The branches were varied as 2, 4 and 8 and all load terminals were terminated in characteristic impedances of the branch. For each case the channel transfer function $H(f)$ was determined using channel transfer function in Ref. [9].

The channel frequency responses were sampled at 1024 and 128 sub-carrier and cyclic prefix, respectively. For the case of noise N_m the square root noise variance in (3) was used. In (4) the values of A and GIR were 0.1 and 0.1, respectively, and m is taken as 3 [8]. Figure 5 shows the performance of the OFDM system modulated with BPSK for various number of branches. It can be observed that to attain

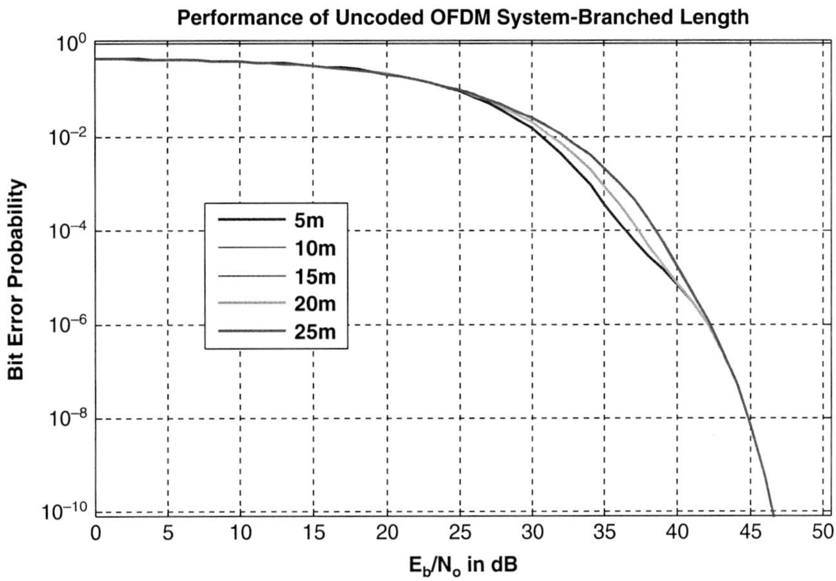

Figure 6: Simulation results for the OFDM system with BPSK modulation for indoor channel for various branch line lengths, with four distributed branches in the link.

a bit error probability of 10^{-10} the SNR per bit of 32, 40 and 58 dB are needed for 2, 4 and 8 branches, respectively. This indicates that the average power needed per branch is about 4 dB/branch so that sustained communication is still maintained.

3.2 Influence of branched line length

To determine the influence of branched line length the same configuration as in Fig. 1 with four branches equally distributed between A and L was considered. The line length between point A and L was 20 m, while the branch lengths were varied as 5, 10, 15, 20 and 25 m. The transmitter and receiver loads were terminated in the line characteristic impedances and the system was assumed to be synchronized while the branched loads terminated in 60 Ω (chosen for having some reflections). Figure 6 shows the performance of the OFDM system modulated with BPSK for various branch lengths. It is seen from Fig. 6 that for indoor channels the influence of branched line length is almost negligible for OFDM systems.

3.3 Influence of load impedances

Again the configuration as in Fig. 1 with four distributed branches was considered with line length between A and L as 20 m and with branch length as 10 m. The branches between point A and L were equally distributed between transmitting

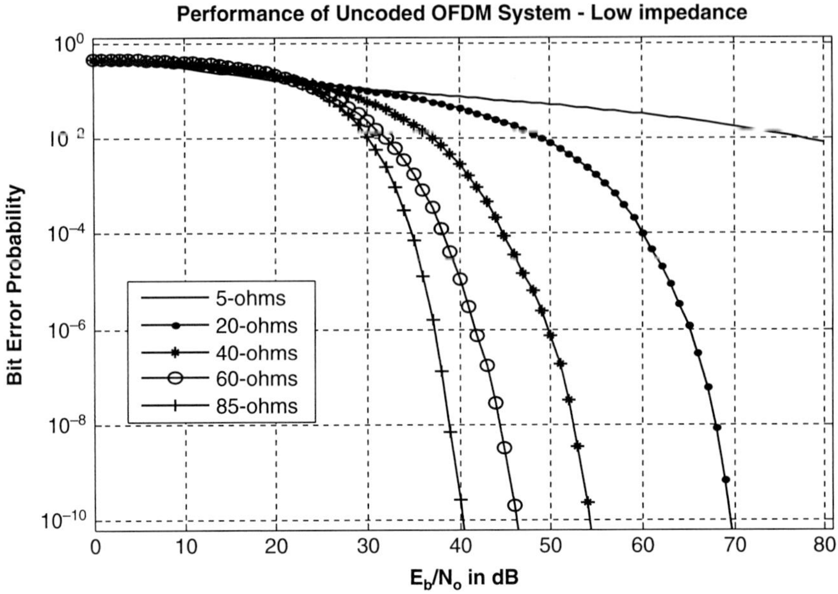

Figure 7: Simulation results for the OFDM system with BPSK modulation for indoor PLC channel for various low load impedances at branch terminations, with four distributed branches in the link.

and receiving ends. The transmitter and receiver loads were terminated in the line characteristic impedances and the system between transmitting and receiving ends was assumed to be synchronized.

3.3.1 Low impedance loads

We consider first the low impedance loads (loads less than branch line characteristic impedance). The load impedances at all terminals were varied as 5, 20, 40, 60 and 85 Ω. Figure 7 shows the performance of the OFDM system with various load impedances. It is observed that the good performance can be obtained when the channel is terminated in characteristic impedances wherein the bit error probability is 10^{-10} at an SNR per bit of 40 dB. When the load impedance decreases by 25 Ω from line characteristic impedance, the power loss is about 0.2 dB/Ω but when the impedance was decreased by 45 Ω, the power loss is about 0.67 dB/Ω. However, as the load impedance approaches a short circuit it leads to a degraded system performance. This is due to the fact that at short circuit, higher deep notches exist in the system.

3.3.2 High impedance loads

We now consider the high impedance loads (impedances higher than the line characteristic impedance). The load impedances at all terminals were varied as 100, 200, 400, 800 and 1600 Ω. Figure 8 shows the performance of the OFDM system

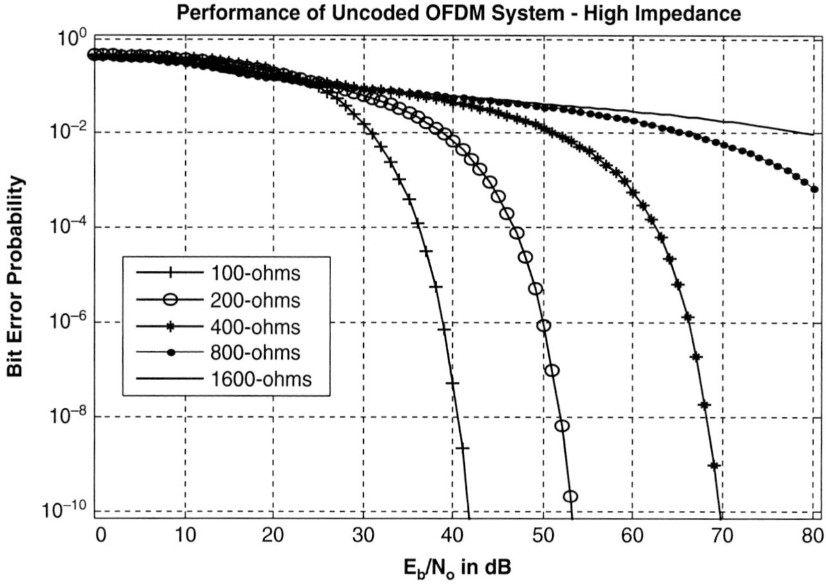

Figure 8: Simulation results for the OFDM system with BPSK modulation for indoor PLC channel for various high impedances at branch terminations, with four distributed branches in the link.

for various impedances. A good channel performance is seen for 100 Ω terminations with the bit error probability of 10^{-10} at an SNR per bit of 42 dB. When the load impedance increases beyond 100 Ω the power loss is about 0.1 dB/Ω and if it increases above 400 Ω the power loss is about 30 dB indicating that performance is getting degraded and as the load impedance approaches open circuit the performance is severely degraded due to deep notches in the system.

4 Performance improvement using concatenated codes

To improve the performance of the BPLC system under load conditions and also for accommodating larger number of branches, concatenated codes as proposed in IEEE802.16a [10] and adopted by Homeplug are considered [11] for the configuration in Fig. 5. The generating functions *T*(*D*) for Viterbi decoder was evaluated using method proposed by Onyszchuk [12], and is given by (6).

$$T(D) = \sum_{d=df}^{\infty} c_d D^d = (B-1-m)T(D,1,1) + \frac{\partial T(D,L,N)}{\partial N} \mid L=1, N=1 \quad (6a)$$

$$T(D,L,N) = \frac{D^5 L^3 N}{1 - DL(1+L)N} \quad (6b)$$

The parameter m in (6) is the constraint length of convolution code which in this chapter is denoted by CC(Rc, m), where Rc denotes code rate. The Reed–Solomon code was represented as RS(n_2, k_2, m). More details of these can be found in Refs. [13, 14]. The system was simulated by considering number of sub-carrier as used in earlier sections. The generating function was compared to get the actual parameters for determination of bit error rate probability for Viterbi encoder.

4.1 Determination of code parameters for system improvement

In all the discussions to follow we consider the degraded performance case of power-line channel with four branches and terminated in 5 Ω (low impedance case) as described in the previous section. For implementation the constraints length in convolution code was taken as 8 with code rate of 1/3. The Reed–Solomon code was considered to correct up to 8, 20, 32 and 64 error bits. Figure 10 shows the performance of concatenated codes for various Reed–Solomon code corrections. The results show that increasing the error correcting capability of the concatenated Reed–Solomon outer code does not always improve the system performance [14]. Better bit error rate (BER) performance is obtained by RS(255, 239) at low values of SNR. Better performance can be obtained also by slightly increasing the error correcting capability, for e.g. see the case for RS(255, 215) as shown in Figs 9 and 10.

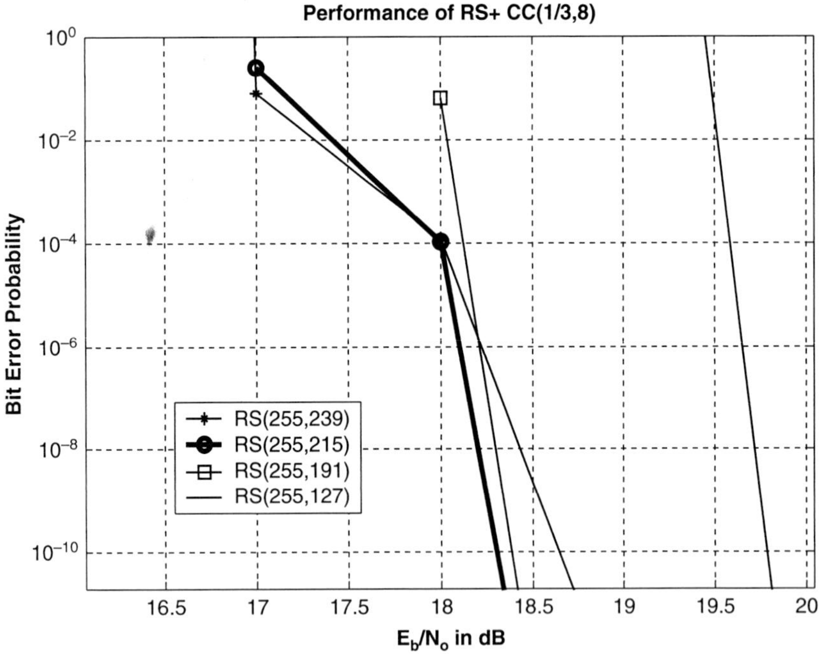

Figure 9: Simulation results for the coded OFDM system with various concatenated Reed–Solomon codes and convolution code for indoor PLC channel terminated in 5 Ω, for IV system.

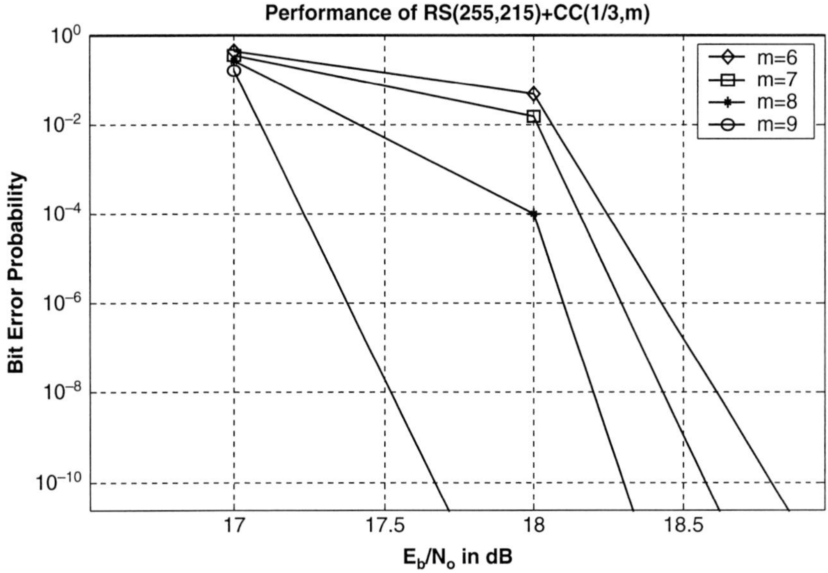

Figure 10: Simulation results for the coded OFDM system with concatenated Reed–Solomon and convolution code (1/3, m) for indoor PLC channel terminated in 5 Ω, for IV system.

Also in comparison to Fig. 7, there is a significant improvement on bit error rate probability (compare the lowest impedance cases).

Now let us consider the case of RS(255, 215, 8) that gave better performance as per Figs 9 and 10. We investigated influence of constraint length on the performance of OFDM system with convolution code rate of 1/3. Figure 11 shows the performance of OFDM system with concatenated codes under various constraint lengths of inner code. It is seen that constraint lengths do not contribute to improvements in system performance (compare Figs 10 and 11). Next let us consider the influence of code rate of inner code on system performance. The code rate for outer Reed–Solomon code was set at RS(255, 215, 8), the system was simulated considering code rate of 1/2, 1/3, 1/5 and 1/7. Figure 11 shows the performance of concatenated codes under various code rate conditions. It is observed that as the code rate increases the performance tends to decrease.

Finally we investigated the performance improvement when different Reed–Solomon codeword lengths are employed with fixed code rate. Figure 12 shows the performance of concatenated Reed–Solomon with CC(1/3, 8). It can be seen that the longer the codeword brings forth minor performance improvement. Note as the codeword length increases the system complexity increases proportionally.

4.2 Performance analysis for OFDM system with concatenated RS(255, 215, 8) and CC(1/2, 8)

In the previous section we analyzed different parameters which can lead to improvement in the performance of a degraded channel using the OFDM system

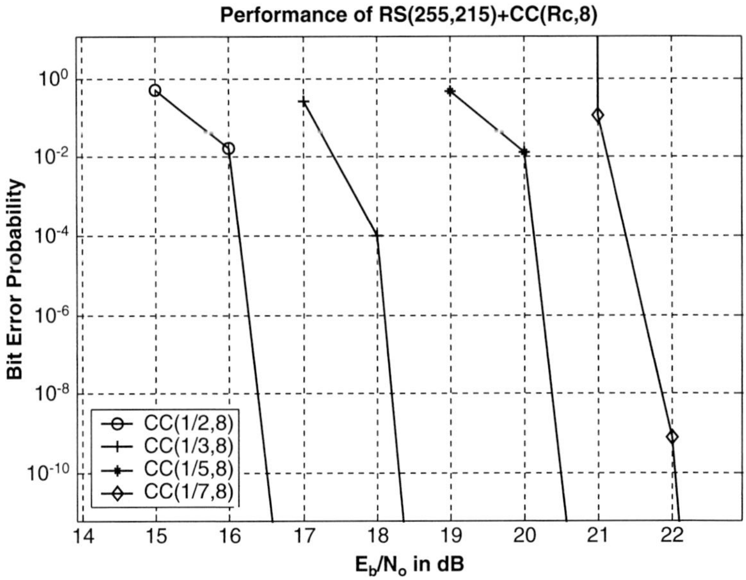

Figure 11: Simulation results for the coded OFDM system with concatenated Reed–Solomon and convolution code (Rc, 8) for indoor PLC channel terminated in 5 Ω.

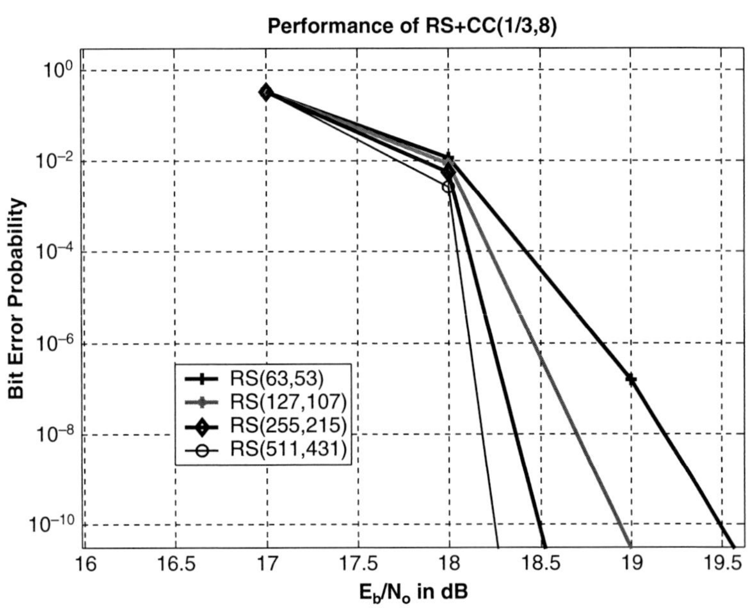

Figure 12: Simulation results for the coded OFDM system with concatenated Reed–Solomon and convolution code (1/3, 8) for indoor PLC channel terminated in 5 Ω.

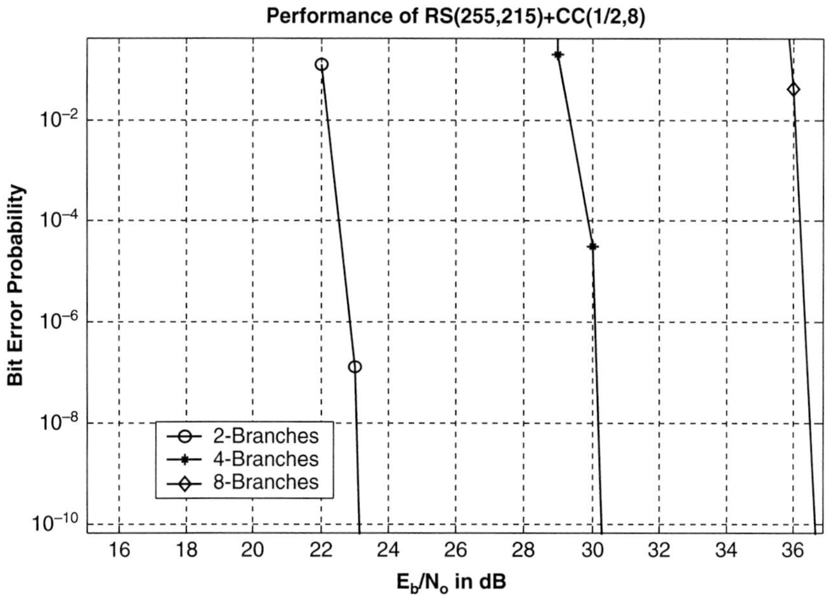

Figure 13: Simulation results for the coded OFDM system with BPSK modulation for indoor PLC channel for various number of branches terminated in characteristic impedances.

with concatenated Reed–Solomon and Viterbi decoding. We shall next apply them for a sensitivity analyses with distributed branches and variable load termination cases as presented earlier.

4.2.1 Influence of number of branches

First let us consider the case with number of branches with simulation conditions for the channel identical to Section 3.3. The parameters used are RS(255, 215, 8) with Viterbi decoder with CC(1/2, 8). Figure 13 shows the performance of concatenated scheme for different number of branches. Comparing the corresponding results with Fig. 5 for an uncoded OFDM system, it is seen that the performance has improved by 10 dB for lower number of branches, but for higher number of branches the concatenated scheme shows an excellent improvement in performance.

4.2.2 Influence of low impedance loads

The simulation conditions are again identical to Section 3.3. The parameters for concatenated scheme are similar to the previous case with number of branches. Figure 14 shows the performance of concatenated Reed–Solomon as outer code and Viterbi as inner code.

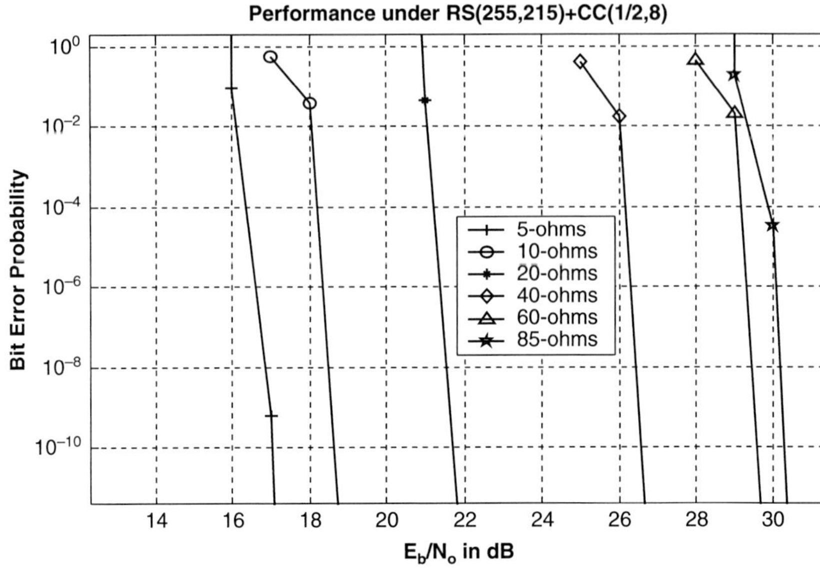

Figure 14: Simulation results for the coded OFDM system with BPSK modulation for indoor PLC channel for various low load impedances at branch terminations.

As expected, it is observed that there is a significant performance improvement for a system terminated in various impedances below the line characteristic impedance, particularly for lower impedances. Compare the corresponding results with Fig. 7 for the uncoded system. For impedances close to characteristic impedance there is an improvement of about 10 dB in comparison to uncoded system.

4.2.3 Influence of high impedance loads

Next let us consider the impedances above the characteristic impedances. The simulation conditions are again identical to those in Section 3.3. The parameters for concatenated scheme are similar to the previous case with number of branches. Figure 15 shows the performance of concatenated Reed–Solomon as outer code and Viterbi as inner code. As expected, it is observed that there is a significant performance improvement for a system terminated in various impedances above the line characteristic impedance, particularly for higher impedances. Compare the corresponding results with Fig. 8 for the uncoded system. For impedances close to characteristic impedance there is an improvement of about 10 dB in comparison to uncoded system.

5 Underground cable systems

Consider a typical underground cable channel used in low voltage (LV) applications [15], whereby the direct line (from transmitter to receiver) is NAYY150SE. The per unit length inductance and capacitance parameters of one of the cable

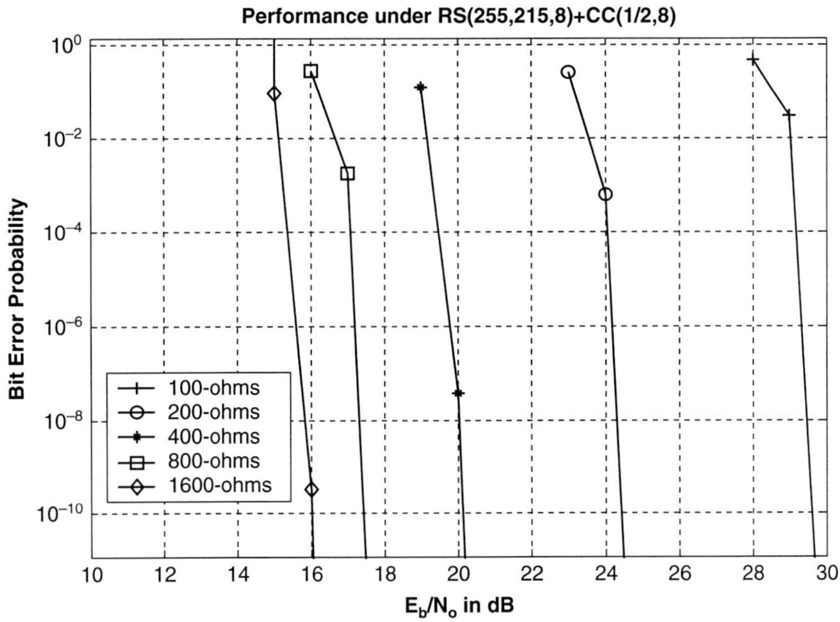

Figure 15: Simulation results for the coded OFDM system with BPSK modulation for indoor PLC channel for various high load impedances at branch terminations.

conductors with adjacent return are 0.32735 μH/m and 0.27191 nF/m, respectively. The branched cable is NAYY35RE with per unit length inductance and capacitance parameters of one of the cable conductors with adjacent return are 0.45179 μH/m and 0.19702 nF/m, respectively. Using the above parameters characteristic impedances of cables NAYY150SE and NAYY35RE are 35 and 48 Ω, respectively [15].

The power-line channel considered here is similar to Fig. 1 with $Z_s = Z_L = 35\ \Omega$. We investigate the variations of direct and branch line lengths and number of branches in the link between A and L. Further variation of branch terminal impedances is also studied. We first determined the channel delay spread, which is a necessary parameter for OFDM multi-carrier system design. After different sensitivity analysis we found that a delay spread for a system with four distributed branches terminated either in 5 Ω or 1.6 kΩ, with direct line length of 1.2 km and branch length of 15 m, constitutes the maximum delay spread case. The maximum delay spread (time span of channel impulse response) T_m is about 4 μs as shown in Fig. 16.

We consider an OFDM system with total frequency band $B = 99.9$ MHz. With such bands, a single-carrier system would have symbol time T_s of 1 ns. Considering T_m of 4 μs, there would be severe ISI. We assumed an OFDM system with BPSK modulation applied to each sub-channel. The channel coherence bandwidth B_c is 0.25 MHz. To ensure flat fading on each sub-channel, we take BN = B/N = 0.1BC. Thus number sub-channels N needed are 3996. In the actual implementations of multi-carrier modulation, N must be a power of 2 for the DFT and IDFT

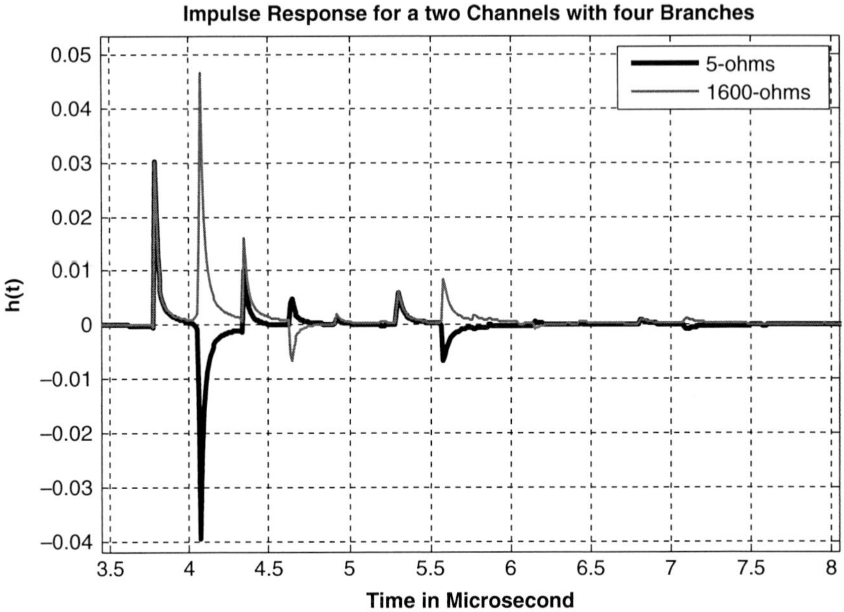

Figure 16: The impulse response of underground power-line channel cables with four branches terminated in 5 and 1600 Ω.

operations, in which case $N = 4096$ is appropriate. So the OFDM symbol duration is TN = N × TS = 40.96 μs. To ensure no ISI between OFDM symbols, the length of cyclic prefix is set to $\mu = 512 > T_m/T_s$; hence, the guard interval $T_{guard} = \mu T_S =$ 5.12 μs. For the case of noise N_m the square root noise variance in (5) was used. In (5) the values of A and GIR were 0.1 and 0.1, respectively, and m is taken as 3 [1]. These design parameters are used in all cases to follow in the sections.

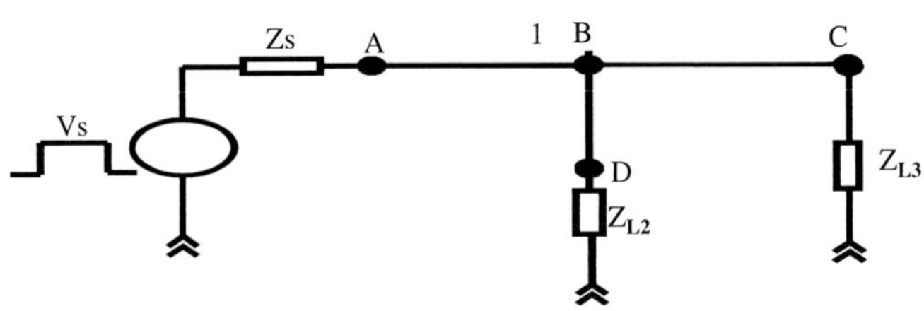

Figure 17: Power-line network with one branch.

5.1 Influence of line length from transmitting point to receiver

To determine the influence of link length from transmitter to receiver, the power-line configuration with one branch as in Fig. 17 was considered. The line length between transmitting and receiving ends AC was varied as 150 m,

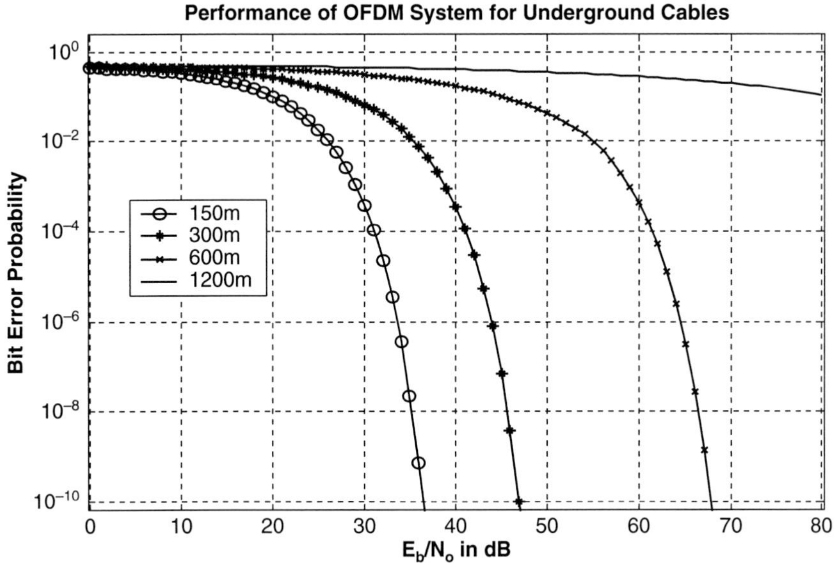

Figure 18: Simulation results for the OFDM system with BPSK modulation for underground cable channel for various direct line lengths.

300 m, 600 m and 1.2 km, and point B was always at the mid point. The transmitter, receiver and branch loads were terminated in the line characteristic impedances and the system was assumed to be synchronized. The branch line length was kept at 15 m.

Figure 18 shows the performance of the OFDM system for various line lengths. It can be observed that to attain a bit error probability of 10^{-10} the SNR per bit of 37, 48, 68 and more than 80 dB are needed for 150 m, 300 m, 600 m, and 1.2 km, respectively. It is found that an average power of 7 dB extra is needed per every 100 m increment in channel length for sustained performance.

5.2 Influence of number of branches

5.2.1 Number of branches distributed at a node

For this, the power-line configuration with distributed branches as in Fig. 19 was considered. The length AC is 400 m and the point B is the midpoint of the direct line and the branches at node B are 15 m long.

The transmitter, receiver and branch loads were terminated in the line characteristic impedances and the system was assumed to be synchronized. The number of branches varied as 2, 4, 8 and 16. Figure 20 shows the performance of the OFDM for various number of branches at a node. It can be observed that to attain a bit error probability of 10^{-10} the SNR per bit of 55, 58 and 61 dB are needed for 2, 4 and 6 branches, respectively. This means an average power of 1.5 dB is needed per every additional branch for a sustained performance.

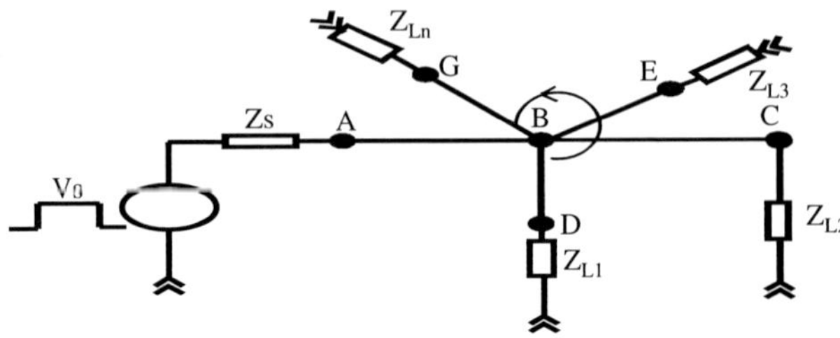

Figure 19: Power-line network with number of branches at a node.

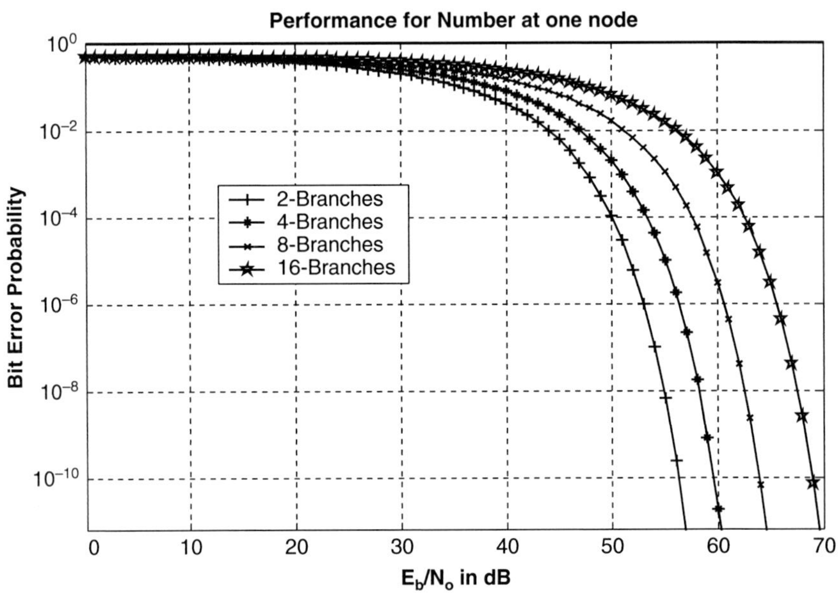

Figure 20: Simulation results for the OFDM system with BPSK modulation for underground cable channel for various numbers of branches at a node.

5.2.2 Number of branches distributed in the link between the transmitter and receiver

For this case the power-line configuration with distributed branches as in Fig. 1 was considered. The branches between point A and L were equally distributed between transmitting and receiving ends. The transmitter and receiver loads were terminated in the line characteristic impedances and the system was assumed to be synchronized. The line length between point A and L was 400 m, while the branch line lengths were kept at 15 m. The branches were varied as 2, 4 and 8, and all branch loads were terminated in characteristic impedances.

Figure 21 shows the performance of the OFDM system for various number of branches. It can be observed that to attain a bit error probability of 10^{-10} the SNR

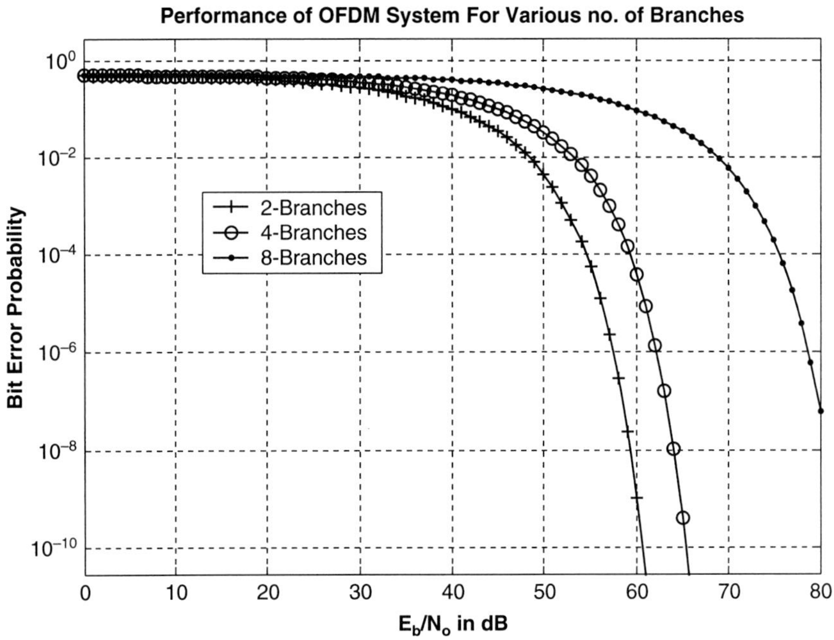

Figure 21: Simulation results for the OFDM system with BPSK modulation for underground cable PLC channel for various number of branches terminated in characteristic impedances.

per bit of 60, 65 and more than 80 dB are needed for 2, 4 and more than 8 branches, respectively. This means that average power needed per branch is about 4 dB/branch for a sustained communication.

5.3 Influence of variation of branch load impedances

Again the configuration with four distributed branches as in previous section was considered but we shall now vary the branch terminal impedances only.

5.3.1 Low impedance branch terminal loads

We consider first the low impedance loads (loads less than branch line characteristic impedance). The branch load impedances were varied as 4, 8, 16 and 34 Ω. Figure 22 shows the performance of the OFDM system with various low load impedance cases. It is observed that the good performance can be obtained when the channel is terminated in characteristic impedances as in previous section. When the load impedance decreases by 23 Ω from line characteristic impedance, the power loss is about 0.35 dB/Ω; but when the impedance decreased below 23 Ω, the power loss is about 0.8 dB/Ω. However, as the load impedance approaches a short circuit it leads to a degraded system performance due to the existence of higher deep notches.

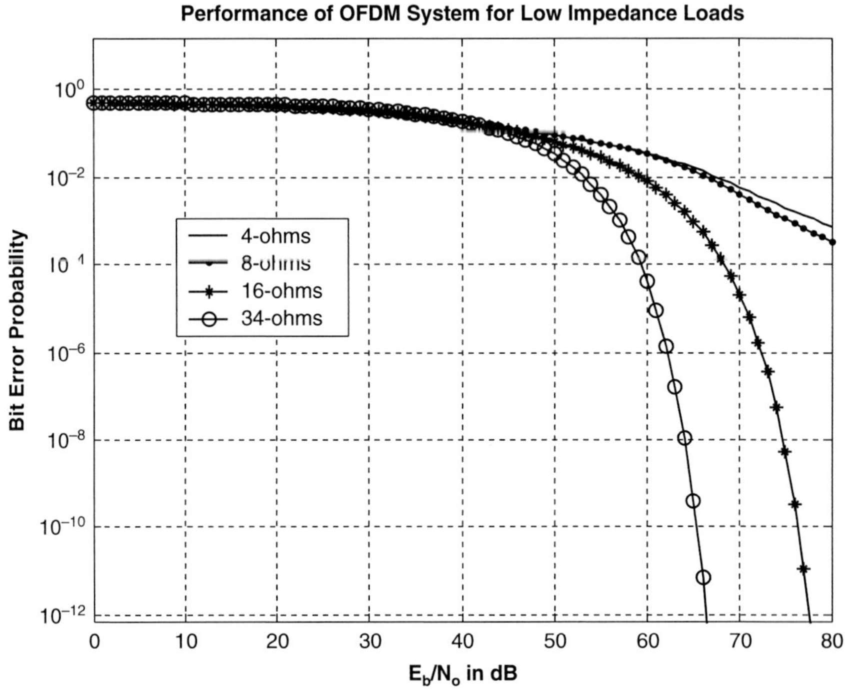

Figure 22: Simulation results for the OFDM system with BPSK modulation for underground cable PLC channel for various low load impedances at branch terminations.

5.3.2 High impedance branch terminal loads

We next consider the high impedance branch loads (impedances higher than the line characteristic impedance). The branch load impedances were varied as 50, 100, 200 and 400 Ω. Figure 23 shows the performance of the OFDM system for various high branch impedance cases. The power is 70 dB, and more than 80 dB, for 50 Ω, and more than 100 Ω, respectively. Above 400 Ω, the power loss is more than 80 dB indicating a degraded performance due to deep notches in the system.

5.4 Influence of branched line length

Here we consider the same system as in previous section with four branches. The branch lengths were varied as 5, 10, 20, 30, 40 and 50 m. The branched loads were terminated in 40 Ω (chosen for having some reflections). Figure 24 shows the performance of the OFDM system modulated with BPSK for various branch lengths. It is seen from Fig. 24 that for underground cable channels, the influence of branched line length is negligible for OFDM systems.

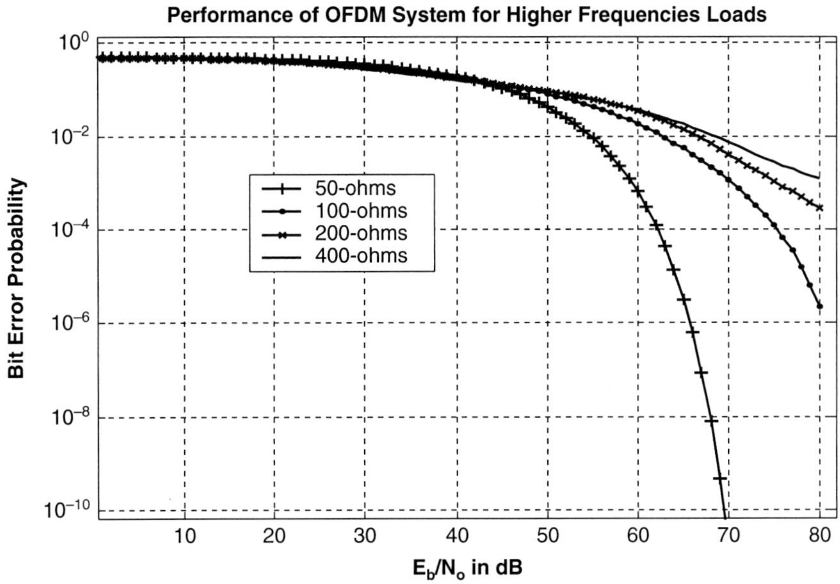

Figure 23: Simulation results for the OFDM system with BPSK modulation for underground cable PLC channel for various high impedances at branch terminations.

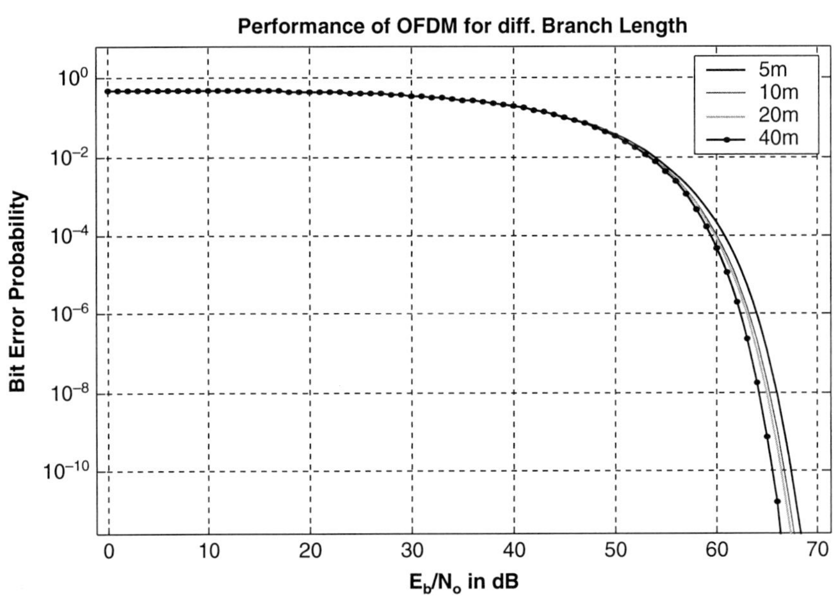

Figure 24: Simulation results for the OFDM system with BPSK modulation for underground cable channel for various branch line lengths.

5.5 Performance improvement using concatenated codes

To improve the performance of the BPLC system under load conditions and also for accommodating larger number of branches, concatenated codes as proposed in IEEE802.16a and adopted by Homeplug are considered. The generating functions *T*(*D*) for Viterbi decoder were evaluated using method proposed by Onyszchuk [12], and is given by (6). The Reed–Solomon code was represented as RS(n_2, k_2, m). The system was simulated by considering number of sub-carrier as used in earlier sections. The generating function in (6) was subjected to soft-decision decoding expression to get the actual parameters for determining bit error rate probability for Viterbi encoder. In the discussions to follow we consider the degraded performance case of power-line channel with four branches and terminated in 400 Ω (high impedance case) as described in earlier section. For implementation the constraints length in convolution code was taken as 8 with code rate of 1/3. The Reed–Solomon code was considered to correct up to 8, 20, 32 and 64 error bits. Figure 25 shows the performance of concatenated codes for various Reed–Solomon code corrections. The results show that increasing the error correcting capability of the concatenated Reed–Solomon outer code does not always improve the system performance. Better BER performance is obtained for either RS(255, 239) or RS(255, 215) for low or high values of SNR.

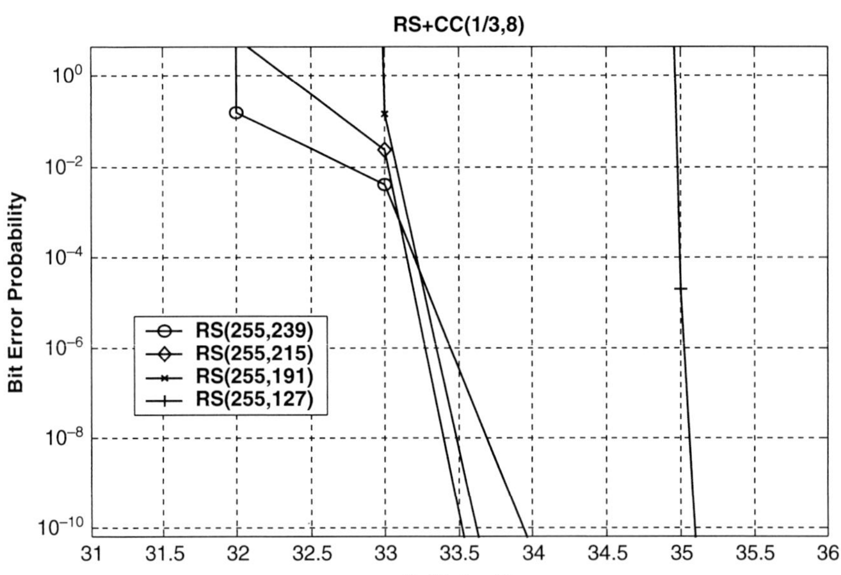

Figure 25: Simulation results for the coded OFDM system with various concatenated Reed–Solomon codes for underground cable PLC channel with branches terminated in 400 Ω with convolution code rate 1/3 and constraint length of 8.

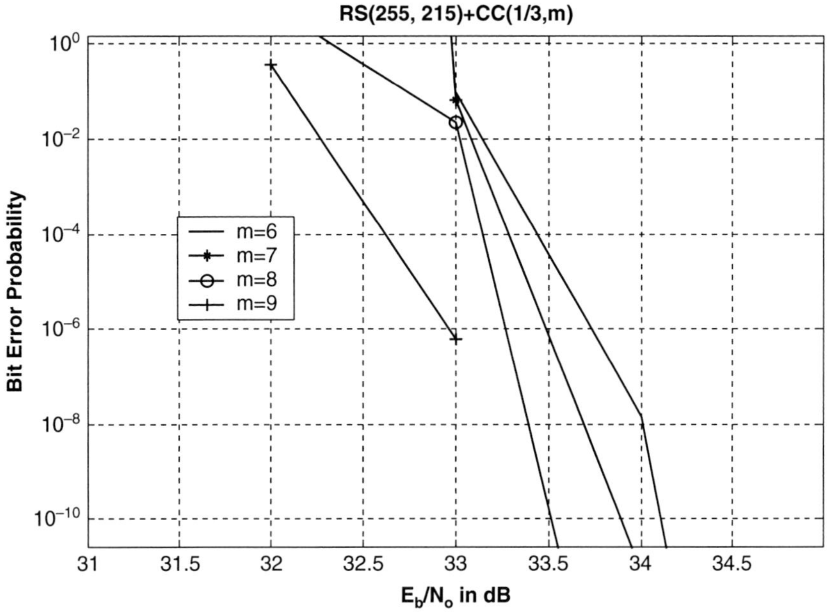

Figure 26: Simulation results for the coded OFDM system with concatenated Reed–Solomon and convolution code (1/3, *m*) for underground PLC channel with branches terminated in 1600 Ω with concatenated Reed–Solomon codes of RS(255, 215).

Considering the case with RS(255, 215, 8) that gave better performance as per Fig. 25, we investigated influence of constraint length on the performance of OFDM system with convolution code rate of 1/3. Figure 26 shows the performance of OFDM system with concatenated codes under various constraint lengths of inner code. It is seen that constraint lengths do not contribute to improvements in system performance (compare Figs 25 and 26). Next let us consider the influence of code rate of inner code on system performance. The code rate for outer Reed–Solomon code was set at RS(255, 215, 8), the system was simulated considering code rate of 1/2, 1/3, 1/5 and 1/7 with $m = 8$. Figure 27 shows the performance of concatenated codes under various code rate conditions. It is observed that as the code rate increases the performance tends to decrease (Fig. 27).

5.6 Performance analysis of OFDM system and concatenated RS(255, 215, 8) and CC(1/2, 8)

In the previous section we analyzed the possible parameters that can lead to improvement in the performance of a degraded channel using the OFDM system with concatenated Reed–Solomon and Viterbi decoding. We shall next apply them for a sensitivity analyses with parameters used are RS(255, 215, 8) with Viterbi decoder with CC(1/2, 8) for distributed branches and variable load termination cases.

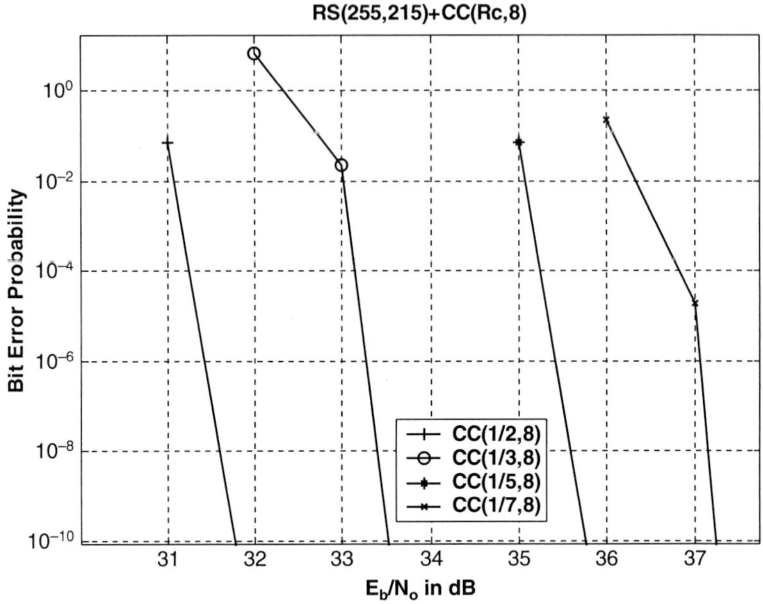

Figure 27: Simulation results for the coded OFDM system with concatenated Reed–Solomon and convolution code (Rc, 8) for underground cable PLC channel with branches terminated in 1600 Ω with concatenated Reed–Solomon codes of RS(255, 215).

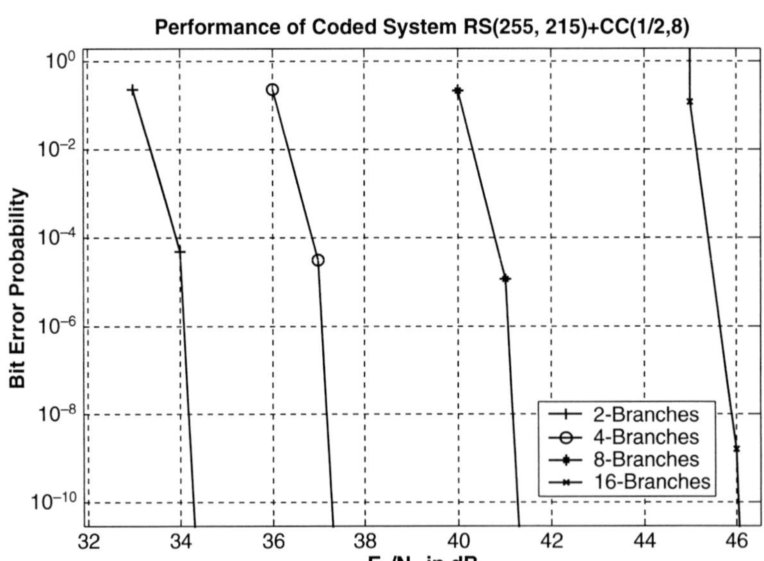

Figure 28: Simulation results for the coded OFDM system with BPSK modulation for underground cable PLC channel for various number of branches concentrated at a node and terminated in characteristic impedances.

5.6.1 Influence of number of branches concentrated at a node

Consider the case with number of branches with simulation conditions for the channel identical to Section 5.2. Figure 28 shows the performance of concatenated scheme for different number of branches. Compare the corresponding results with Fig. 20 for an uncoded OFDM system, it is seen that the performance has improved by more than 20 dB.

Consider the case with number of branches with simulation conditions for the channel identical to Section 5.2.1. Figure 29 shows the performance of concatenated scheme for different number of branches. Comparing Figs 21 and 29, it is seen that the performance has improved by 10 dB for lower number of branches but for higher number of branches the concatenated scheme shows an excellent improvement in performance.

5.6.2 Influence of number of branches distributed in the link between the transmitter and receiver

5.6.2.1 Influence of low branch terminal impedance loads

Consider the simulation conditions as in Section 5.3. Figure 30 shows the performance-coded OFDM system. It is observed that there is a significant performance improvement for a system terminated in various low impedances below the line characteristic impedance, particularly for lower impedances (compare Figs 22 and 30). For impedances close to characteristic impedance there is an improvement of about 20 dB in comparison to uncoded system.

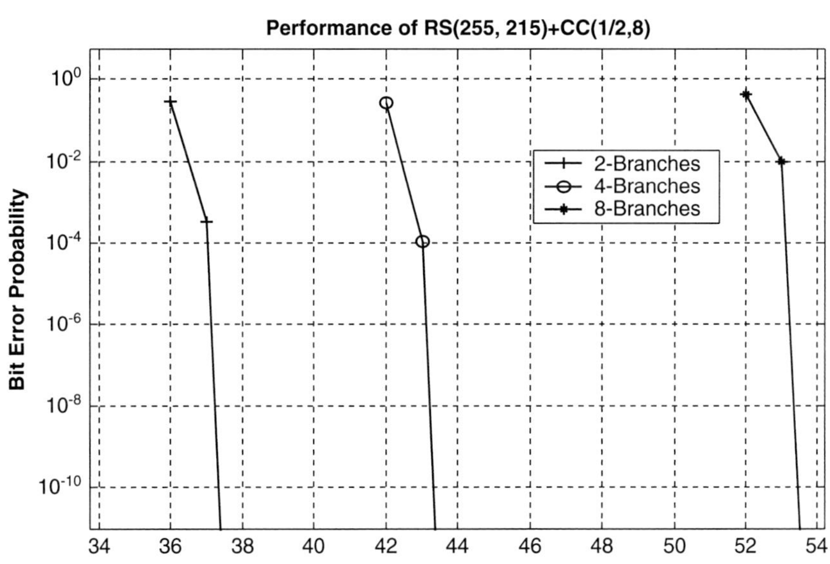

Figure 29: Simulation results for the coded OFDM system with BPSK modulation for underground PLC channel for various number of branches distributed in the link between the transmitter and receiver and terminated in characteristic impedances.

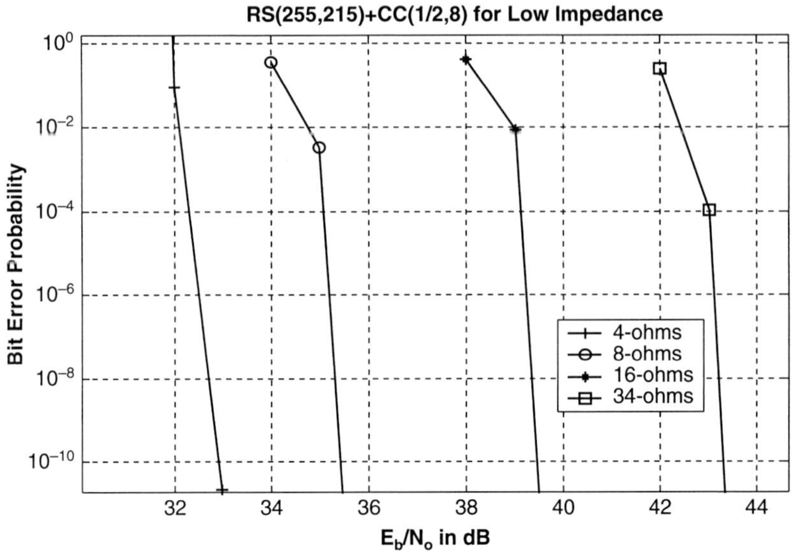

Figure 30: Simulation results for the coded OFDM system with BPSK modulation for underground cable PLC channel for various low load impedances at branch terminations.

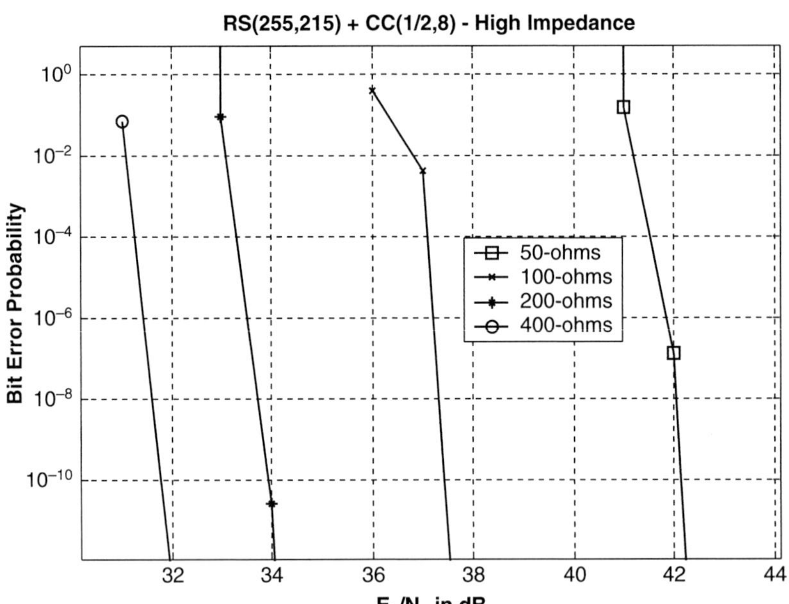

Figure 31: Simulation results for the coded OFDM system with BPSK modulation for underground cables PLC channel for various high load impedances at branch terminations.

5.6.2.2 Influence of high terminal impedance loads

The simulation conditions are identical to those in Section 5.3. Figure 31 shows the performance of the coded OFDM system. There is a significant performance improvement for a system terminated in various impedances above the line characteristic impedance, particularly for higher impedances (compare Figs 23 and 31). For impedances close to characteristic impedance, there is an improvement of about 10 dB in comparison to the uncoded system.

References

[1] Fukami, T., Umehara, D., Kawai, M. & Morihiro, Y., Noncoherent PSK Optimum Receiver over Impulsive Noise Channels. *International Symposium of Information Theory and Its Applications*, Xi'an, PRC, pp. 235–238, October 2002.

[2] Middleton, D., Statistical–physical model of electromagnetic interference. *IEEE Transactions on Electromagnetic Compatability*, **19(3)**, pp. 106–126, August 1977.

[3] Proakis, J.G., *Digital Communications*, McGraw-Hill International Edition, Fourth Edition, 2001.

[4] Amirshashi, P., Navidpour, S.M. & Kavehrad, M., Performance analysis of uncoded and coded OFDM broadband transmission over low voltage powerline channels with impulsive noise. *IEEE Transactions on Power Delivery*, **21(4)**, October 2006.

[5] Zander, A.J. & Slimane, B., *Principles of Wireless Communications*, Narayana Press: Denmark, 2006.

[6] Goldsmith, A., *Wireless Communications*, Cambridge University Press, 2005.

[7] Prasad, R., *OFDM for Wireless Communications Systems*, Artech House: Boston, MA, 2004.

[8] Anatory, J., Theethayi, N. & Thottappillil, R., Effects of multipath on OFDM systems for indoor broadband power-line communication networks, *IEEE Transactions on Power Delivery*, pp. 1190–1197, **24(3)**, July 2009.

[9] Anatory, J., Theethayi, N. & Thottappillil, R., Power-line communication channel model for interconnected networks – part I: two conductor system. *IEEE Transactions on Power Delivery*, **24(1)**, January 2009.

[10] Huang, K.L. & Hang, H.M., Performance Analysis for Serially Concatenated FEC in IEEE802.16a over Wireless Channels. *Pacific Rim Conference on Multimedia*, Tokyo, Japan, Lecture Notes in Computer Science 3332; SCI, IF: 0.513, December 2004.

[11] Lee, M.K., Newman, R., Latchman, H.A., Katar, S. & Yonge, L. HomePlug 1.0 powerline communication LANs – protocol description and comparative performance results. *Special Issue of the International Journal on Communication Systems on Power-Line Communications*. pp. 447–473, May 2003.

[12] Onyszchuk, I., Finding the complete path and weight enumerators of convolutional codes. *TDA Progress Report*, JPL, pp. 203–213, February 1990.

[13] Cideciyan, R.D., Eleftheriou, E. & Rupf, M., Concatenated Reed–Solomon/convolutional coding for data transmission in CDMA-based cellular systems. *IEEE Transactions on Communications*, **45(10)**, pp. 1291–1303, October 1997.

[14] Wong, Y.F.M. & Letaief, K.B., Concatenated coding for DS/CDMA transmission with wireless communications. *IEEE Transactions on Communications*, **48(12)**, pp. 1965–1969, December 2000.

[15] Anatory, J., Theethayi, N., Thottappillil, R., Kissaka, M.M. & Mvungi, N.H., The influence of load impedance, line length and branches on underground cable power-line communications (PLC) systems. *IEEE Transactions on Power Delivery*, **23(1)**, pp.180–187, January 2008.

Index